Reviews of Physiology, Biochemistry and Pharmacology 148

Springer

Berlin
Heidelberg
New York
Hong Kong
London
Milan
Paris
Tokyo

Reviews of

148 Physiology Biochemistry and Pharmacology

With 20 Figures and 3 Tables

Springer

ISSN 0303-4240
ISBN 3-540-40136-9 Springer-Verlag Berlin Heidelberg New York

Library of Congress-Catalog-Card Number 74-3674

Springer-Verlag Berlin Heidelberg New York
a member of BertelsmannSpringer Science+Business Media GmbH

http://www.springer.de

Printed in Germany

Printed on acid-free paper – 27/3150 ag 5 4 3 2 1 0

Rev Physiol Biochem Pharmacol (2003) 148:1–80
DOI 10.1007/s10254-003-0009-x

F. Wehner · H. Olsen · H. Tinel · E. Kinne-Saffran · R. K. H. Kinne

Cell volume regulation: osmolytes, osmolyte transport, and signal transduction

Published online: 4 April 2003

Abstract In recent years, it has become evident that the volume of a given cell is an important factor not only in defining its intracellular osmolality and its shape, but also in defining other cellular functions, such as transepithelial transport, cell migration, cell growth, cell death, and the regulation of intracellular metabolism. In addition, besides inorganic osmolytes, the existence of organic osmolytes in cells has been discovered. Osmolyte transport systems—channels and carriers alike—have been identified and characterized at a molecular level and also, to a certain extent, the intracellular signals regulating osmolyte movements across the plasma membrane. The current review reflects these developments and focuses on the contributions of inorganic and organic osmolytes and their transport systems in regulatory volume increase (RVI) and regulatory volume decrease (RVD) in a variety of cells. Furthermore, the current knowledge on signal transduction in volume regulation is compiled, revealing an astonishing diversity in transport systems, as well as of regulatory signals. The information available indicates the existence of intricate spatial and temporal networks that control cell volume and that we are just beginning to be able to investigate and to understand.

General introduction

For a number of years, cell volume regulation in mammalian cells has been considered of minor physiological importance since overall osmolality of the plasma is one of the very tightly regulated parameters of the body. In the meantime, however, it has been realized

E. Kinne-Saffran deceased on December 6, 2002

F. Wehner (✉) · H. Olsen · E. Kinne-Saffran · R. K. H. Kinne
Max-Planck-Institut für molekulare Physiologie, Otto-Hahn-Str. 11,
44227 Dortmund, Germany
e-mail: frank.wehner@mpi-dortmund.mpg.de · Tel.: +49-231-1332225

H. Tinel
Institute of Cardiovascular Research II, Bayer AG Wuppertal, 42096 Wuppertal, Germany

that quite a variety of tissues such as renal medullary cells and chondrocytes are exposed to anisotonic extracellular media and thus require volume regulatory mechanisms. Anisotonic conditions can also arise in pathological conditions such as hypo- or hypernatraemia when homeostatic functions of the body are insufficient (Law 1999; Law 1998; Verbalis et al. 1989; Verbalis 1994). Similarly, cells encased in a rigid surrounding, such as brain cells, depend on cell volume regulation for their proper function. Furthermore, even at normal extracellular osmolalities cells can generate transmembrane osmotic differences due to uptake of (organic) osmolytes or during transepithelial transport of solutes (McCarty and O'Neil 1991).

Cell volume changes also occur globally during maturation of erythrocytes, cell growth, differentiation, hypertrophy, and apoptosis. Locally, cell migration and shape changes require volume adaptations (Lang et al. 1998b). Finally, cell volume has been identified as a mechanism that regulates cell metabolism (Lang et al. 1989). This effect is particularly evident in the liver, where cell swelling increases protein and glycogen synthesis and cell shrinkage increases protein and glycogen breakdown (Häussinger 1996; Häussinger 1998). Thus, cell volume and its importance during the whole life cycle of a cell has become more and more evident and this area of physiology and pathophysiology has been attracting an increasing number of investigators.

During recent years, it also became more and more accepted that not only inorganic osmolytes such as K^+ and Cl^- are employed to restore osmotic equilibrium across the cell membranes but that also organic osmolytes play a significant role in cell volume homeostasis.

In most reviews, inorganic and organic osmolytes and their respective volume regulatory transporters are dealt with separately. The current review combines the two and discusses the question of a putative identity of some systems. This contribution also takes into account the progress made in identifying the molecular entities of the transporters by molecular biology techniques. Furthermore, the current state of knowledge with regard to the regulation and coordination of the various osmolyte transporters is reviewed in order to shed some light on the intriguing signal transduction networks used in cell volume regulation.

The osmometric behavior of cells

Any changes in cell volume of animal cells under anisotonic conditions are, in principle, based on a distinct permeability of the plasma membrane to water. In many instances, this permeability is increased by specialized membrane proteins, which mediate the facilitated diffusion of water, so-called water channels or aquaporins (Maunsbach et al. 1997; Marinelli and LaRusso 1997; Ma and Verkman 1999). Given a sizeable water permeability at its outer membrane, a cell then exhibits an "osmometric" behavior as its first passive response to anisotonicity, i.e., a movement of water will occur, which changes cell volume until the difference in osmotic pressure is equalized and a new chemical equilibrium of water across the plasma membrane is achieved. From a physicochemical point of view, however, a cell is not a *perfect* osmometer, which means that its volume will not change to the same extent as is predicted by the law of Boyle/van't Hoff. This means that apparently some 20%–40% of total cell volume usually comprises a phase that is not "cytosolic," which means that it is not osmotically active (Lucke and McCutcheon 1932). In practice, the osmotically active/inactive space of a living cell can be easily determined graphi-

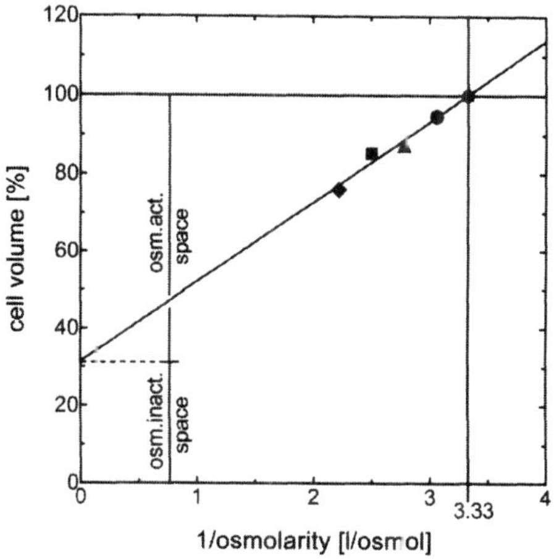

Fig. 1 With all Na^+ importers of RVI blocked (i.e., with 10^{-3} mol/l amiloride plus 10^{-4} mol/l furosemid; see text for details), the *passive* changes of rat hepatocyte volumes after 10 min exposure to hypertonicity (327, 360, 400, and 450 mOsm/l) are a linear function of the reciprocal of extracellular osmolarity. The value of 3.33 is equivalent to 300 mOsm/l where cell volumes equal 100% (control, ★). The intercept of the regression line with the volume axis is equivalent to the "osmotically inactive space" that amounts to 31.3% of total cell volumes (from Wehner and Tinel 2000)

cally by plotting its relative volume versus the reciprocal of extracellular osmolarity, which is equivalent to a Boyle/van't Hoff-Plot (see Fig. 1).

The intercept of the regression line with the Y-axis then gives the osmotically inactive space; its difference from the 100% mark is equivalent to the osmotically active part. This type of analysis, which deals with the passive osmotic behavior of a cell, however, is only valid if the active compensatory processes by which cells readjust their volume are not activated and/or are inhibited efficiently. In rat hepatocytes, for instance, this was achieved by blocking all Na^+ import mechanisms mediating regulatory volume increase (RVI) (namely Na^+ conductance, Na^+/H^+ antiport, and Na^+-K^+-$2Cl^-$ symport) by 10^{-3} Mol/l amiloride and 10^{-4} Mol/l furosemide and by then determining the actual changes of cell volume after 10 min of cell shrinkage, under four different hypertonic conditions. From these measurements, an osmotically inactive space could be determined that amounts to some 31% of total cell volume (Wehner and Tinel 2000).

Inorganic and organic osmolytes

For a number of years, research on volume regulation in mammalian cells has focused on the movement of the main intracellular inorganic osmolytes such as potassium and chloride. Compensatory changes in these osmolytes form indeed the backbone for volume regulation when only small perturbations of the intra- or extracellular osmolarity occur.

In recent years it has become, however, more and more evident that—similar to other phyla in nature—in mammalian cells also organic osmolytes are employed in volume regulation (Yancey et al. 1982; Hoffmann et al. 1988; Kinne 1993; Burg 1995; Burg 1996; Junankar and Kirk 2000).

Organic osmolytes comprise polyols such as sorbitol and myo-inositol, methylamines such as glycerophosphorylcholine and betaine, and amino acids such as taurine, glutamic acid, and ß-alanine amongst others (Yancey and Burg 1989; Yancey et al. 1982; Beck et al. 1985; Beck et al. 1998).

There are several hypotheses as to why in some cells organic solutes are used in addition to inorganic solutes. One hypothesis is the "compatible osmolyte hypothesis" which is based on the observation that high concentrations of inorganic salts such as NaCl or KCl perturb the function of enzymes or other proteins (Noulin et al. 1999), whereas organic solutes do not (Yancey et al. 1982). Another hypothesis reflects the "counteracting osmolyte principle," which refers to the finding that methylamines attenuate the destabilizing effect of high concentrations of urea on protein function (Burg 1996). A third aspect is that several organic osmolytes are electroneutral and can replace inorganic osmolytes which, when released across the cell membrane, may change the membrane potential and thereby neuronal excitability in the brain (Iwasa et al. 1980) or driving forces for electrogenic sodium-cotransport systems. Again, the brain is particularly interesting because of the existence of reuptake systems for excitatory amino acids that are transported by sodium-cotransport systems (Curtis and Johnston 1974).

One of the major places for organic osmolyte accumulation is the renal medulla, where accumulation occurs because of a broad range of extracellular osmolalities exceeding normal osmolality (for review see Beck et al. 1985; Bagnasco et al. 1986; Yancey and Burg 1989; Garcia-Perez and Burg 1991; Kinne et al. 1993; Kinne 1993; Kinne et al. 1995; Kinne 1998; Grunewald and Kinne 1999; Kinne et al. 2001) in particular in the direction of hyperosmolality (Grunewald et al. 1993a; Grunewald et al. 1994; Handler and Kwon 2001). Also chondrocytes (de Angelis et al. 1999; Hall and Bush 2001) encounter hyperosmolality in the extracellular space because of the high concentration of fixed charges in the mucopolysaccharides in which they are embedded. Limited expandability in the brain (Strange 1992; Pasantes-Morales et al. 1994a; Pasantes-Morales et al. 1994b; Pasantes-Morales et al. 2000b) and regulation of cell transparency in the lens (lens fiber cells; Burg and Kador 1988; Cammarata et al. 2002), also require organic osmolytes.

Intracellular accumulation of organic osmolytes involves mainly two processes. The first represents uptake across the cell membrane by specific transport systems, the second intracellular generation of the osmolyte by metabolic reactions. Examples of the former are sodium-cotransport systems for myo-inositol, taurine, and betaine which use the sodium gradient across the plasma membrane as driving force. The general and specific properties of these cotransport systems are summarized in the section entitled "Organic osmolytes in RVI."

Metabolic reactions are primarily responsible for the generation of sorbitol from D-glucose and glycerophosphorylcholine (GPC). D-glucose taken up by the cells via uniport or symport systems can be readily converted into sorbitol by the enzyme aldose reductase using NADPH as cofactor (Gabbay 1973; Ohta et al. 1990). In an extensive study on renal inner medullary aldose reductase, the affinity for D-glucose was found to be about 370 mMol/l, the affinity for NADPH was 7.5 μMol/l. Thus, in the intact cell with a NADPH concentration of about 0.4 mMol/l the enzyme is saturated with NADPH, but only to a small extent with D-glucose. Due to the high K_m of the enzyme, in the intact cell sorbitol synthesis is characterized by the affinity of the cellular glucose uptake systems. For example, a value of about 60 mMol/l is obtained in intact IMCD cells, which reflects the K_m usually observed for glucose uniporters GLUT 1 or GLUT 2 (Grunewald et al. 1993b). Regulation of the synthesis occurs mainly by altering the V_{max} of the enzyme; a strong increase is observed in a variety of cells when they are exposed to hypertonic conditions in vivo or in vitro (Bedford et al. 1987; Bagnasco et al. 1988; Grunewald et al. 1998; Lang et al. 1998a). However, in disease states such as diabetes, the NADPH/NADP ratio is another important regulator of the rate of sorbitol synthesis (Grunewald et al. 1993b).

Sorbitol breakdown yielding fructose can occur via the sorbitol dehydrogenase pathway with NAD as the hydrogen acceptor. In some tissues that use sorbitol as organic osmolyte, these pathways are restricted to different, sometimes closely associated cells (Grunewald et al. 1995; Kinne et al. 1997). In general, the activity of sorbitol dehydrogenase is much lower than that of aldose reductase, thereby limiting the importance of this enzyme in osmotic adaptation (Grunewald et al. 1995; Kinne et al. 1997; Grunewald et al. 1998).

The major pathways involved in the metabolism of GPC in IMCD cells, for example, are the following. The precursor of GPC, choline, is taken up by sodium-independent transport systems. Choline is phosphorylated and then incorporated into the phospholipid phosphatidylcholine (PC). By a stepwise removal of the fatty acid residues (involving phospholipase A2), first lysophosphatidylcholine and then GPC are generated. Interestingly, this reaction seems to use a pool of PC different from that present in the plasma membrane. Breakdown of GPC is mediated by the GPC: choline diesterase reaction yielding choline and phosphoglycerol. The latter two reactions seem to be slow compared to the synthetic pathway. The predominant "osmosensitive" enzyme is the GPC: choline diesterase since studies in vivo and in vitro indicate that after longterm exposure of cells to hypertonic conditions, the rate of enzymatic breakdown is reduced and thus the overall concentration of the organic osmolyte is increased (Zablocki et al. 1991; Bauernschmitt and Kinne 1993).

The release of organic osmolytes from the cells involves channel-like proteins, in some instances organic and inorganic osmolytes share the same transporter. Thus, a complex picture emerges in which inorganic and organic osmolyte levels have to be controlled in a well-coordinated manner (Burg 1996). This coordination requires feedback systems that have to be elucidated and taken into account when considering volume regulatory processes in the cells.

Regulatory volume increase

Role of inorganic osmolytes in RVI

In order to achieve RVI, the intracellular osmolyte content has to be augmented rapidly. To this end, transport systems for inorganic osmolytes are activated as the first cellular response. The main osmolyte taken up by the cells is sodium since favorable driving forces for this cation exist across the plasma membrane. Sodium is subsequently exchanged against potassium by the action of the Na^+,K^+-ATPase, to restore the original sodium gradients and its electrochemical potential.

There are also mechanisms set in place that reduce loss of intracellular potassium and sometimes also chloride, depending on its electro-chemical equilibrium.

Na^+/H^+ antiport

Na^+/H^+ antiporters (NHEs) catalyze the secondary active and electroneutral exchange of H^+ against Na^+. With one exception, they are most important for the regulation of cell pH, but some NHEs are effective mediators of RVI. Six NHE isoforms have been cloned to date (NHE1–NHE6) which exhibit a common molecular organization (see for review

Orlowski and Grinstein 1997; Counillon and Pouyssegur 2000; Ritter et al. 2001). They consist of two main functional domains with amino acid identities of 45%–60% and 25%–35%, respectively. These are an N-terminal portion of 10–12 transmembrane regions (depending on the hydropathy algorithm used) and a large intracellular C-terminal part that is equipped with regulatory sites (see below). NHE1 to NHE5 are located in the plasma membrane, while NHE6 is sorted to the mitochondrial inner membrane. NHE1 and NHE6 are ubiquitously expressed whereas NHE2–NHE5 are restricted to specific tissues. In epithelia, NHE1 appears to be largely confined to the basolateral membrane (Coupaye-Gerard et al. 1996) but, in cell lines like OK (opossum kidney) or MDCK (Madin-Darby canine kidney), it may also occur apically (Noel et al. 1996). NHE2, NHE3, and NHE4 are found preferentially in the gastrointestinal tract and kidney where they appear to reside mainly in the apical membrane (Soleimani et al. 1994; Dudeja et al. 1996; Noel et al. 1996; Hoogerwerf et al. 1996; Sun et al. 1997; Bookstein et al. 1997). The presence of NHE2 in the kidney and NHE4 in the intestine, however, is discussed controversially (Bookstein et al. 1997). NHE5 was detected in nonepithelial tissues, preferentially in the brain, but also in spleen, testis, and skeletal muscle (Klanke et al. 1995). In many cells, several NHE isoforms are expressed and assigned to specific membrane domains so that (together with other transporters) they achieve a concerted cross-talk in the vectorial movement of ions; this especially holds for epithelia (Ritter et al. 2001).

Some NHEs are inhibited by submillimolar concentrations of amiloride and, more selectively, by its derivatives dimethyl-amiloride and ethyl-isopropyl-amiloride (Tse et al. 1994; Counillon and Pouyssegur 2000). There are significant differences in sensitivity to these blockers among the different isoforms, which (for amiloride) amount to almost four orders of magnitude. The amiloride sensitivity generally follows the order NHE1 > NHE2 >> NHE3 > NHE4 (Orlowski and Grinstein 1997; the latter two isoforms are therefore sometimes referred to as being amiloride-insensitive (Counillon and Pouyssegur 2000). The mitochondrial isoform NHE6 exhibits a rather low sensitivity to amiloride, but it is efficiently blocked by the analog, benzamil (Brierley et al. 1989).

With respect to their role in cell volume regulation, it was found that NHE1, NHE2, and NHE4 are activated under hypertonic conditions, whereas NHE3 is inhibited (Orlowski and Grinstein 1997; Hoffmann and Mills 1999; Ritter et al. 2001). NHE1 is clearly the Na^+/H^+ antiport most commonly employed in RVI, which is not surprising in regard to the ubiquitous expression of the protein. As a matter of fact, the shrinkage-induced activation of Na^+/H^+ antiport that had been reported in a variety of preparations prior to the molecular definition of NHE isoforms can now in many instances be attributed to NHE1 (Lang et al. 1998a; Lang et al. 1998b; Hoffmann and Mills 1999).

Concerning regulation, hypertonicity appears to shift the pH_i sensitivity of NHE1 to more alkaline values (Grinstein et al. 1985). Interestingly, although the C-terminal domain of NHE1 exhibits a number of phosphorylation sites (that are actually used for the activation by hormones and mitogens), no direct phosphorylation of the protein occurs upon cell shrinkage (Sardet et al. 1990; Sardet et al. 1991; Grinstein et al. 1992). Rather, there appears to be a significant hypertonicity-induced cross-talk between NHE1, stress fibers and the small GTPase Rho so that osmotically-induced changes in the actin texture (also involving integrins) are very likely to be part of the signaling cascade-mediating activation (Watson et al. 1992; Grinstein et al. 1993; Vexler et al. 1996; Hooley et al. 1996; Tominaga et al. 1998; Tominaga and Barber 1998). In addition, Ca^{2+} may participate in the stimulation because the C-terminal domain of NHE1 contains two calmodulin binding sites and deletion or mutation of the high affinity binding site significantly reduces the osmotic sensitivity of the transporter (Bertrand et al. 1994). It is worth mentioning that NHE1 is also subject to transcriptional osmotic regulation (Ritter et al. 2001).

The hypertonic activation patterns of NHE2 are less well-defined. Proline-rich regions in the C-terminal portion of the protein appear to be involved which resemble SH3 binding domains and might reflect sites of interaction with the cytoskeleton and/or signaling molecules. Truncation experiments, however, revealed that these regions are probably involved in the proper targeting of NHE2 rather than its regulation (Chow et al. 1999). Besides this, NHE2 is subject to transcriptional control (Soleimani et al. 1994; Bai et al. 1999).

NHE3 is inhibited under hypertonic conditions. This effect is due to a reduction of the maximal transport velocity rather than to changes in substrate affinities (Nath et al. 1996). NHE3 is highly sensitive to the actin texture of cells (Kurashima et al. 1999; Szaszi et al. 2000). The protein associates with two regulatory factors, namely NHERF1 and NHERF2, and NHERF1 could be shown to bind to ezrin, an actin-binding protein of the ezrin-radixin-moesin (ERM) family that links membrane proteins to the cytoskeleton. Based on deletion experiments, it could also be shown that the same C-terminal region of NHE3 mediates actin sensitivity and NHERF binding (Kurashima et al. 1999).

Moreover, among the small GTPases of the Rho family controlling actin assembly, the inhibitory form of RhoA (but not Rac1 and Cdc42) greatly depressed NHE3 activity and comparable effects were observed upon a specific block of Rho kinase (ROK), a downstream effector of RhoA; furthermore, inhibition of ROK reduced the phosphorylation of myosin light chain (MLC). These data strongly suggest that the RhoA-ROK signalling pathway is a mechanism for the control of NHE3 activity, which is, at least in part, achieved by controlling the phosphorylation of MLC and, consequently, the organization of the actin cytoskeleton (Szászi et al. 2000).

Interestingly, considerable amounts of NHE3 are present in recycling endosomes (Ritter et al. 2001). This predominantly intracellular location may contribute to its paradoxical behavior during RVI.

NHE4 is clearly activated under hypertonic conditions; it exhibits a bell-shaped profile with maximal functionality close to 500 mOsm/l (Bookstein et al. 1994). This relatively low osmotic sensitivity is readily explained in terms of NHE4 tissue distribution with maximal expression found in the renal medulla (Bookstein et al. 1994; Ritter et al. 2001) where a high extracellular osmolality prevails.

Na^+-K^+-$2Cl^-$ symport

NKCC1 and NKCC2 are mediators of electroneutral Na^+, K^+, Cl^- cotransport across cell membranes at a stoichiometry of 1:1:2 (with very few exceptions; see Haas and Forbush, III 2000; Russell 2000 for reviews). On the molecular level, they share many similarities with other members of Cl^--dependent cation transporters, namely the Na^+-Cl^- symporter (NCC) and the four isoforms of K^+-Cl^- symporters (KCC1 to KCC4) cloned so far. These similarities include molecular weights in the range of 110 kD to 130 kD (deglycosylated), 12 predicted transmembrane regions, and large hydrophilic intracellular N- and C-terminal domains. The most conserved regions of these transporters are the transmembrane domains, as well as the putative intracellular loops connecting them, particularly the one between TM2 and TM3. NKCCs are blocked by the "loop" diuretics bumetanide and furosemide at micromolar concentrations; KCCs exhibit a lower affinity to these compounds, whereas a possible inhibition of NCC remains ambiguous (Russell 2000). NKCC1 exhibits a broad tissue distribution and is found in many secretory epithelia (where it resides in the basolateral membrane), as well as in a variety of nonepithelial cells. In contrast, NKCC2 is only present (apically) in the thick ascending limb of Henle's loop and in the macula densa of the kidney (Haas and Forbush, III 2000; Russell 2000).

NKCC1 is one of the most routinely employed transporters of cell volume regulation (Russell 2000) during RVI (Lang et al. 1998a; Lang et al. 1998b; Hoffmann and Mills 1999; Haas and Forbush, III 2000). Activation of NKCC1 appears to involve phosphorylation at serine/threonine residues and cell shrinkage results in NKCC1 phosphorylation in a number of cells. In most instances, however, the actual kinase in charge remains to be identified (Hoffmann and Mills 1999; Haas and Forbush, III 2000). Recently, the phosphorylation of NKCC1 by c-Jun NH_2-terminal kinase (JNK), a member of the mitogen-activated protein kinase (MAPK) family, was reported from bovine aortic endothelial cells that were shrunken under hypertonic or isotonic conditions (Klein et al. 1999). In rat hepatocytes, the hypertonic activation of Na^+-K^+-$2Cl^-$ symport (as well as that of Na^+ conductance) was inhibited by staurosporine, as well as by the PKC specific blocker bis-indolylmaleimide I (BIM; Heinzinger et al. 2001). In human tracheal epithelial cells, the hypertonic activation of NKCC1 appeared to be mediated by PKC-δ; this process is also likely to involve the (extracellular signal-regulated) kinase ERK (Liedtke and Cole 2002). In addition to phosphorylation and dephosphorylation, the activity of NKCC1 appears to be regulated by the state of the actin network, as well as by accessory proteins that remain to be characterized (Haas and Forbush, III 2000; Russell 2000).

Cation channels

Compared to the symport system discussed above, the rates of ion transport by channels are some 4–5 orders of magnitude higher. Accordingly, any modulation of channel activity in response to changes of cell volume will serve as a fast and very efficient regulatory mechanism.

If one considers the electrochemical driving forces for Na^+, K^+, and Cl^- transport across most cell membranes, the activation of Na^+-selective channels and conductive Na^+ entry would be a highly efficient mediator of RVI. The resultant depolarization of membrane voltage would favor a parallel conductive entry of Cl^-. Na^+ accumulating inside the cell would then be extruded via Na^+, K^+-ATPase so that constant driving forces for Na^+-coupled cotransporters are ensured. In sum, a net intracellular gain of K^+ and Cl^- and thus a rapid increase of cell volume would be achieved by these mechanisms.

There is an increasing number of systems from which the hypertonic activation of cation channels is reported. Two main classes of channels can be distinguished based on their sensitivity to amiloride. In the following, first the amiloride sensitive channels will be discussed.

The activation of conductive Na^+ entry as a mechanism of RVI was originally proposed by Hoffmann (Hoffmann 1978) and Okada (Okada and Hazama 1989) for Ehrlich ascites tumor cells and the Intestine 407 cell line, respectively. In 1995, a hypertonic stimulation of cell membrane Na^+ conductance was reported from current-clamp recordings on rat hepatocytes in confluent monolayer culture (Wehner et al. 1995; see Fig. 2).

Furthermore, in a quantitative study, it could be shown that the relative contribution of Na^+ conductance, Na^+/H^+ antiport, and Na^+-K^+-$2Cl^-$ symport to the initial uptake of Na^+ under hypertonic stress was approximately 4:1:1 (Wehner and Tinel 1998). This clearly

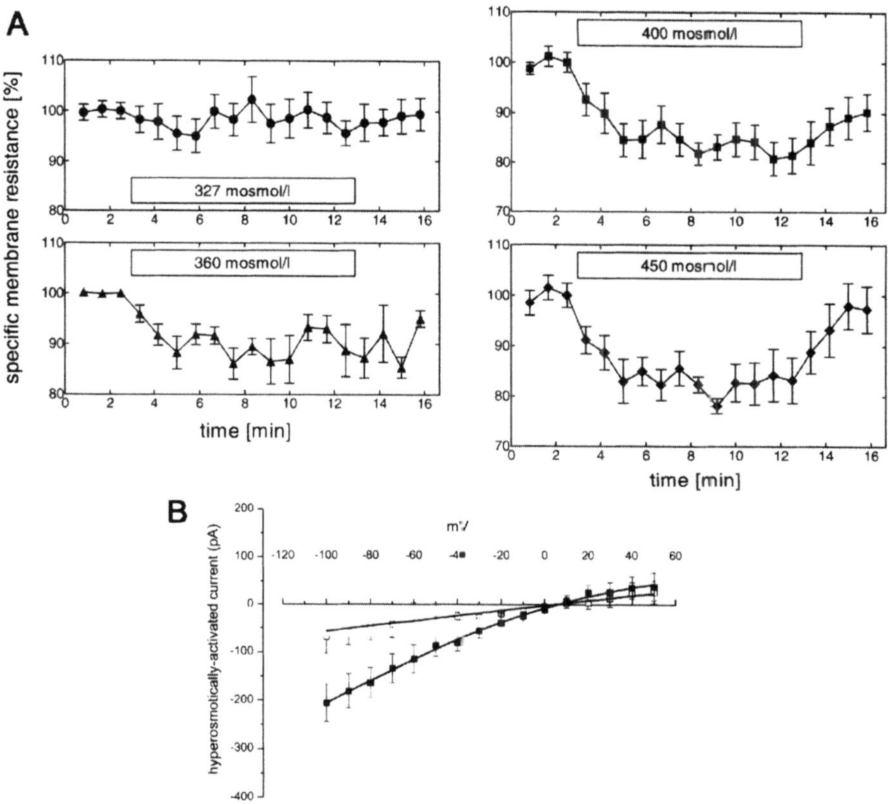

Fig. 2 **A** Cable analysis of specific membrane resistance, reflecting the hypertonic activation of Na^+ conductance in rat hepatocytes. Experiments were carried out in the continuous presence of 0.5 mM quinine; extracellular osmolarity was increased from 300 mOsm/l to the osmolarities indicated; means±SEM; n=4–5. **B** Current-voltage relationships of hypertonicity-induced membrane currents in rat hepatocytes obtained with (two-channel microelectrode) voltage-clamp techniques. The differences between currents obtained at 400 mOsm/l and 300 mOsm/l are depicted for cells injected with control-oligo-DNA (■) or anti-α-rENaC oligo DNA (□) (from Wehner and Tinel 2000; Böhmer and Wehner 2001)

renders Na^+ conductance the prominent mechanism of RVI in this system. The Na^+ conductance was inhibited by amiloride with an apparent K_i of 6 μMol/l and its overall sensitivity profile was EIPA > amiloride > benzamil (Wehner et al. 1997; Böhmer et al. 2000). Hence, at first sight, the hypertonicity-induced Na^+ conductance of rat hepatocytes may reflect a low (amiloride) affinity type rather than an epithelial Na^+ channel (ENaC); the latter typically exhibits K_i values in the upper nanomolar range and an inverse pharmacological profile (Garty and Palmer 1997; Fyfe et al. 1998). In voltage-clamp experiments on rat hepatocytes, a relatively low cation selectivity of the channel was detected with a P_{Na}/P_K of 1.4 (Böhmer and Wehner 2001). In patch-clamp experiments in the cell-attached configuration, hypertonicity-induced single channel events with a unitary conductance of 6 pS were recorded (P. Lawonn and F. Wehner unpublished observation). With respect to intracellular signaling events, the hypertonic activation of Na^+ conductance (and Na^+-K^+-$2Cl^-$ symport, but not Na^+/H^+ antiport) is mediated by PKC (Heinzinger et al. 2001).

In the human hepatoma cell-line HepG2, hypertonic stress led to the activation of a nonselective cation conductance that was clearly sensitive to amiloride with 65% inhibition at 10^{-5} Mol/l (Wehner et al. 2002a). Interestingly, this hypertonicity-induced conductance was also partially inhibited by Gd^{3+} (10^{-4} Mol/l) and flufenamate (10^{-5} Mol/l) which are typical blockers of nonselective but amiloride-insensitive cation channels in a variety of systems (see below). Benzamil and EIPA (at 10^{-5} Mol/l each) had no effect on HepG2 cation conductance (Wehner et al. 2002a). In Ehrlich-Lettre ascites tumor cells, hypertonic conditions activated a cation conductance that did not discriminate between Na^+, K^+, and Li^+, but that was impermeable to NMDG and choline. In the range of 10^{-6} Mol/l to 10^{-3} Mol/l this cation conductance was blocked by the ion channel inhibitors benzamil > Gd^{3+} > amiloride > EIPA (in order of potency; Lawonn et al. 2003). In immortalized human nonpigmented ciliary epithelial cells, shrinkage upon return from hypo-osmotic to normosmotic conditions was followed by a partial (post-RVD) RVI (Civan et al. 1996) that was mediated by the parallel operation of Na^+/H^+ antiport, Na^+-Cl^- symport, Na^+-K^+-$2Cl^-$ symport, as well as by conductive Na^+ entry. In the range of 10^{-6} Mol/l and 10^{-5} Mol/l, this Na^+ conductance was significantly inhibited by amiloride and—even more efficiently—by benzamil, the most effective blocker of the ENaC (Civan et al. 1997; Garty and Palmer 1997; Fyfe et al. 1998). In principal cells of rat cortical collecting duct hypertonic stress led to a distinct depolarization of membrane voltage coinciding with an increase of cell Na^+ (Schlatter et al. 1997). These effects were partially inhibited by 10^{-5} Mol/l amiloride or Gd^{3+} and the effects were additive. This result was interpreted in terms of a parallel activation of Na^+ channels and nonselective cation channels.

In human red blood cell ghosts, hypertonic stress induced a cation conductance that appeared to be equally permeable to Na^+ and K^+ (but impermeable to NMDG) and that was partially inhibited by 10^{-4} Mol/l amiloride (Huber et al. 2001). These findings support earlier flux studies on lamprey erythrocytes in which hypertonic stress led to the activation of an amiloride-sensitive Na^+ transport that could not be attributed to Na^+/H^+ or Na^+/Na^+ antiport (Gusev and Sherstobitov 1996).

In U937 macrophages, hypertonic stress led to the activation of an inward current with a distinct selectivity for Na^+ over K^+ that was inhibited by amiloride with an apparent K_i close to 1 μMol/l. At the single channel level, this cation conductance appeared to be related to a 6 pS channel. Interestingly, these channels are upregulated by pretreatment of the cells with glucocorticoids, which are known to regulate macrophage function (Gamper et al. 2000).

The shrinkage-activated amiloride sensitive cation channel appears to be related to (α)ENaC. In rat hepatocytes, antisense nucleotides attenuated the activation of the channel (Böhmer and Wehner 2001; see Fig. 2B). In human ciliary body, α-ENaC mRNA was identified (Civan et al. 1997). Similarly, mRNA for α, β, and γ subunits was detected in rat hepatocytes and these subunits could also be identified on the protein level (Böhmer and Wehner 2001). Although the nonselectivity of hypertonicity-induced channels is quite in contrast to the high P_{Na}/P_K reported for α-, β-, γ-rENaC (which is as high as 40; Fyfe et al. 1998) this is not contradictory per se to a possible contribution of ENaCs because their biophysical properties (also including single-channel conductance) strongly depend on subunit composition (Garty and Palmer 1997) as well as their relative affinity to amiloride and its derivatives (Benos et al. 1997; Garty and Palmer 1997). Furthermore osmo-sensitivity of α-, β-, γ-ENaC was reported when the channel subunits were expressed in oocytes (Awayda and Subramanyam 1998). Since some of the results are, however, conflicting, further studies are necessary to clarify the relation between ENaC and the amiloride-sensitive cation channels.

The second group of hypertonicity-induced cation channels is clearly *amiloride-insensitive* up to concentrations of 10^{-4} Mol/l. Quite typically, however, these channels are inhibited by the anti-inflammatory drug flufenamate (10^{-4} Mol/l), as well as by Gd^{3+} in the range of 10^{-5} to 10^{-3} Mol/l. They are expressed in human nasal epithelial cells (Chan and Nelson 1992), in the human colon cell-lines CaCo-2 and HT_{29} (Nelson et al. 1996; Koch and Korbmacher 1999), in the mouse cortical collecting duct cell line M1 (Volk et al. 1995), as well as in BSC-1 renal epithelial cells (derived from the African green monkey), A 10 vascular smooth muscle cells (established from rat embryonic aorta), and Neuro-2a cells (derived from mouse neuroblastoma; Koch and Korbmacher 1999). In general, these channels are rather nonselective with respect to monovalent cations, although there are certain peculiarities. In human nasal epithelium and in CaCo-2 cells, for instance, channels did not discriminate much between Na^+, K^+, and Cs^+, but exhibited a significantly lower permeability to Li^+ and, in the latter system, there was also a distinct permeability to NMDG (P_{Na}/P_{NMDG}=0.56; Chan and Nelson 1992; Nelson et al. 1996). In contrast, in the M1 cell-line the channel appeared to be equally permeable to Na^+, K^+, Cs^+, Li^+, and Rb^+ but virtually impermeable to NMDG (Korbmacher et al. 1995; Volk et al. 1995). In addition, there appear to be significant differences in the relative permeabilities of amiloride-insensitive channels to Cl^- with P_{Na}/P_{Cl} values in the range of 60 (Korbmacher et al. 1995) to 1.7 (Nelson et al. 1996). In some systems, the hypertonic activation of channels is Ca^{2+}-independent (Chan and Nelson 1992) whereas, in others, channel activity appears to require a minimum $(Ca^{2+})_i$ of 1 μMol/l (Korbmacher et al. 1995; Koch and Korbmacher 1999). Nevertheless, these channels do not appear to conduct significant amounts of divalent cations (Korbmacher et al. 1995; Koch and Korbmacher 2000). The unitary conductances of hypertonicity-induced but amiloride-insensitive channels generally is in the range of 15 pS to 27 pS (Korbmacher et al. 1995; Koch and Korbmacher 1999).

Most interestingly, at least some of these amiloride-insensitive nonselective cation channels are typically inhibited by cytosolic ATP concentrations in the millimolar range (Koch and Korbmacher 1999). This raises some concerns as to their actual role in RVI because, under most physiological conditions these channels will remain silent. Also of note in this respect, in a recent study on isolated rat colonic crypt cells, there was no detectable activation of nonselective cation channels under hypertonic stress (Weyand et al. 1998), whereas these cells are known to express Ca^{2+}-activated channels which become detectable in excized inside-out patches and which are very similar to those observed in HT29 cells (Bleich et al. 1996). Clearly, whereas nonselective and amiloride-insensitive cation channels appear to be ubiquitously expressed, their actual contribution to cell volume regulation remains to be elucidated.

In principle, the activation of cation conductances—either sensitive or insensitive to amiloride—that do not discriminate much between Na^+ and K^+ will lead to both Na^+ influx as well as K^+ efflux. Because of the inside negative membrane voltage, however, overall cation gain will significantly exceed overall cation loss equivalent to an increase of the overall intracellular osmotic activity.

K^+ channels

In most systems, the continuous channel-mediated K^+ leakage out of cells is equivalent to a significant loss of cellular osmolytes. This K^+ loss is commonly compensated for by the activation of Na^+, K^+-ATPase. Nevertheless, it is rather obvious that any inhibition of K^+ channels under hypertonic conditions will per se facilitate RVI and, in some cells, this mechanism appears to be an important mediator of volume regulation.

An inhibition of K^+ conductance was originally reported for the basolateral membrane of toad and rabbit urinary bladder (Lewis et al. 1985; Donaldson et al. 1989), as well as of frog skin (Costa et al. 1987; Leibowich et al. 1988) where cells were shrunken by extracellular Cl^- removal (leading to a loss of Cl^- because of the lower membrane permeability of the Cl^- substitute used when compared to Cl^- itself) and/or by increasing extracellular osmolarity. In addition, hypertonic stress reduced basolateral K^+ conductance in rabbit proximal tubule (Lapointe et al. 1990; Macri et al. 1997) and human nasal epithelium (Willumsen et al. 1994) and it decreased conductive K^+ loss in MDCK cells (Ritter et al. 1991). In mouse liver, a reduction of cell membrane K^+ conductance appeared to be the main mechanism of RVI (Graf et al. 1988; Wang and Wondergem 1991) and a transient inhibition of K^+ channels may contribute to the volume response of rat hepatocytes (Wehner et al. 1995; Wehner and Tinel 1998). Interestingly, whole-cell recordings on freshly isolated hippocampal neurons revealed a decrease of voltage-gated K^+ currents under hypertonic conditions (Huang and Somjen 1997). In dissociated rabbit corneal epithelial cells, a large conductance K^+ channel (167 pS in symmetrical 150 mMol/l KCl) could be identified that appeared to mediate the decrease in whole-cell K^+ conductance to 44% of control upon change to 130% extracellular osmolarity (Farrugia and Rae 1993). In a recent study on isolated rat colonic crypts, it was found that hypertonic stress (+50 mOsm/l and +100 mOsm/l) led to membrane depolarizations by 12 mV and 22 mV, respectively, coinciding with decreases in whole-cell conductance to 70% and 50% of control (Weyand et al. 1998). On the molecular level, these effects appeared to be correlated to a 16 pS K^+ channel that experienced decreases in activity ($N \cdot P_o$) to 47% and 44% when compared to isotonic conditions. Of note, this reduction of channel activity was most likely triggered by a hypertonicity-induced decrease of $(Ca^{2+})_i$. The channel was inhibited by (10^{-4} Mol/l) Ba^{2+} and (10^{-9} Mol/l) charybdotoxin (Bleich et al. 1996).

Anion channels

Inhibition of anion channels during RVI might directly contribute to the gain in osmolytes if the intracellular anion (Cl^- or HCO_3^-) activity is above electrochemical equilibrium. Indirectly, it restricts the movement of the counterion for K^+ flux across the membrane, thus impeding K^+ losses.

Decreases of Cl^- conductance under hypertonic conditions have been reported only from a limited number of preparations. In cultured human nasal epithelium, hypertonic stress reduced apical Cl^- conductance (Willumsen et al. 1994) whereas, in rabbit collapsed proximal convoluted tubules, the partial Cl^- conductance of the basolateral membrane was inhibited (Macri et al. 1997). In some instances, the decrease of Cl^- conductance under hypertonic conditions may reflect an inhibition of hypotonicity-induced Cl^- channel activation. In human vas deferens cells in primary culture, for example, with 290 mOsm/l solutions in the experimental bath, as well as in the patch pipette, there was a slowly developing increase of Cl^- conductance once the whole-cell configuration was achieved; this effect could be reversed by an increase of extracellular osmolarity (Winpenny et al. 1996). The increase of Cl^- conductance was interpreted to be due to the additional osmotic activity of intracellular macromolecules and the resultant swelling of cells.

Organic osmolytes in RVI

In most cells investigated, RVI after exposure to a hypertonic extracellular medium occurs within minutes and involves the net uptake of inorganic osmolytes. During this short-term regulation of cell volume, changes in organic osmolytes transport are generally not involved, although it has been observed that the plasma membrane permeability for organic osmolytes such as sorbitol (e.g., in IMCD cells grown at 600 mOsm/l and exposed to 900 mOsm/l medium) decreases (Bagnasco et al. 1988; Grunewald and Kinne 1989). Thus, leak pathways for organic osmolytes are downregulated to make the intracellular accumulation of organic osmolytes—by transporters or metabolic synthesis—more effective. Whether these leak pathways are identical to the ones activated during RVD remains to be investigated.

If exposure of cells to hypertonicity is extended to periods of hours or days, adaptive changes take place aimed to replace the augmented intracellular inorganic electrolytes by organic osmolytes. To this end, intracellular synthesis is increased (sorbitol and GPC) or intracellular breakdown is decreased (GPC; for review see Burg 1995; Burg et al. 1997). For taurine, betaine, and myo-inositol, the predominant effect is on the rate of uptake, which increases severalfold in a matter of hours (see Fig. 3).

Thus, strictly speaking, organic osmolytes are not directly involved in RVI; they serve, however, in the long run to maintain cellular electrolyte homeostasis. Therefore, the cellular uptake systems for organic osmolytes and their regulation are described in the following paragraphs.

Na^+-Cl^--taurine symport

Taurine, a β-amino acid with a sulfonic acid instead of a carboxylic acid as head group occurs in plasma at concentrations of 40 µMol/l. It has several major functions: as partner in the formation of taurocholate, as neurotransmitter (Jacobsen and Smith 1968; Curtis and Johnston 1974), and as an organic osmolyte. Although several tissues, including liver and astrocytes, can synthesize taurine from cysteine, the high intracellular concentration (e.g., 20 mMol/l in IMCD cells) can only be achieved by an active uptake into the cell (Ruhfus et al. 1998). The transporter mediating this uptake is a secondary active sodium-chloride taurine cotransport system in which up to three sodium ions are translocated with one chloride ion and one taurine molecule. Taurine transport with these characteristics has been found in renal (Chesney et al. 1985a, Kinne et al. 1998; Wolff and Kinne 1988; Zelikovic et al. 1989; Zelikovic and Budreau-Patters 1999) and intestinal brush border (O'Flaherty et al. 1997), in liver cells (Warskulat et al. 1997a, Warskulat et al. 1997b; Peters-Regehr et al. 1999), in H4IIE hepatoma cells (Warskulat et al. 1997b), Ehrlich ascites tumor cells (Hoffmann et al. 1988), astrocytes (Beetsch and Olson 1996), the retinal pigment epithelium (Handler and Kwon 2001), bovine aortic endothelial cells (Qian et al. 2000), renal IMCD cells, and MDCK cells (Uchida et al. 1991). The apparent affinity

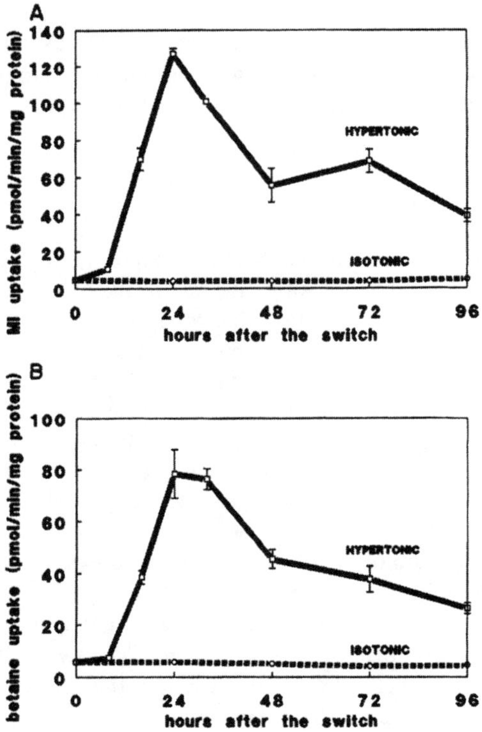

Fig. 3 Time course of basolateral *myo*-inositol (**A**) and betaine (**B**) uptake into MDCK cells switched into hypertonic medium. On day 0, cells cultured on filters in defined medium with 10% FBS were switched to same medium made hypertonic (500 mOsm/l) by addition of raffinose. Isotonic cells were maintained in isotonic defined medium with 10% FBS. Uptake was performed at 37°C for 30 min with 10 μMol/l *myo*-inositol and 10 μMol/l betaine. Results are means±SD to triplicate samples (from Yamauchi et al. 1991)

(K_m) for taurine is in the range of 10 μM to 100 μM (Uchida et al. 1991). The transporter also transports β-alanine and, with a lower affinity, γ-amino butyric acid (GABA). Molecular cloning from MDCK cells revealed that the transporter corresponds to a protein with 655 amino acids with a relative molecular mass of 74 kDa and probably has twelve transmembrane helices (Uchida et al. 1992). It shows significant amino acid sequence similarity to the MDCK cell betaine/GABA transporter and other Na^+, Cl^--dependent neurotransmitter transporters (Guastella et al. 1990). In dog tissues, the order of mRNA abundance for this transporter (dTAUT) was kidney cortex ≈ kidney medulla > ileal mucosa > brain > liver > heart > epididymis. Taurine transporters cloned from other sources such as LLC-PK1 cells from pig kidney, human retinal pigment epithelium, thyroid, placenta, and bovine aortic endothelial cells (Qian et al. 2000) were of similar size (621 and 620 amino acids, respectively) and have a high amino acid sequence similarity.

In a variety of cells and tissues such as MDCK cells (Uchida et al. 1991), human and bovine lens epithelial cells (Cammarata et al. 2002), bovine aortic endothelial cells (Qian et al. 2000), rat liver macrophages (Warskulat et al. 1997a), H4IIE hepatoma cells (Warskulat et al. 1997b), primary cultures of rat hepatocytes (Warskulat et al. 1997a), rat astrocyte cultures (Beetsch and Olson 1996), and human Caco-2 cells (Satsu et al. 1999), exposure to hyperosmotic media results in an increase in V_{max} of taurine transport and a concomitant increase in transporter mRNA (Burg et al. 1997; Handler and Kwon 2001; Cammarata et al. 2002).

High affinity taurine transport is inhibited by phosphorylation via PKC by decreasing affinity for taurine and sodium and reducing maximal velocity (Kulanthaivel et al. 1991;

Brandsch et al. 1993; Nakamura et al. 1996; Mollerup and Lambert 1998; Lima et al. 2000; Qian et al. 2000). Furthermore, taurine uptake adaptively decreases when taurine availability is augmented (Chesney et al. 1985b; Han et al. 1998). Although these regulatory processes appear not to be linked to the response to hypertonicity, they complicate the study of the molecular mechanisms of the latter. However, it can be assumed that the mechanisms involved in osmotic adaptation are similar to the ones described below for the betaine transport system, since both systems respond in a very comparable manner.

Na^+-Cl^--betaine symport

In medullary kidney cells and chondrocytes, which use betaine in volume regulation, intracellular concentrations of up to 50 mMol/l are found. Although cells usually contain choline dehydrogenase that catalyses the synthesis of betaine from choline, most of the betaine is taken up from the extracellular medium. The same holds for MDCK (Yamauchi et al. 1991) and PAP-HT25 cells (Ferraris et al. 1996). The uptake of betaine is dependent on the presence of both sodium and chloride and the transport system (Yamauchi et al. 1992; Moeckel et al. 1997) has been identified to belong to the family of brain GABA and noradrenaline transporters (Guastella et al. 1990).

Initial cloning of the transporter from MDCK cells revealed a protein of 614 amino acids with a molecular weight of 69 kDa. Its proposed membrane topology predicts 12 transmembrane helices and an intracellular NH_2 and COOH terminus (Yamauchi et al. 1992). When expressed in oocytes, the protein showed both betaine and γ-amino-n-butyric acid (GABA) transport activities that were chloride- as well as sodium-dependent (Matskevitch et al. 1999; Forlani et al. 2001). Because of these transport properties and the 43%–49% identity in the amino acid sequence (Rasola et al. 1995) with the rat brain GABA transporter and the human brain noradrenaline transporter, it was named betaine-GABA-transporter (BGT1).

Consecutive studies showed that the gene extends over 28 kilobases and consists of 18 exons. In addition, the 5′ end of the gene has three different first exons, thus a complex mixture of mRNAs exists. Three main types of mRNA (A, B, and C) can be distinguished that differ considerably in their 5′ untranslated sequences. Each type is expressed in a tissue-specific manner: kidney medulla (and MDCK cells) contain all three types, brain and kidney cortex, type B and C, and liver only type A (Takenaka et al. 1995; Burg et al. 1997).

Betaine transport shows apparent affinities (K_m) of 0.3 mMol/l–0 5 mMol/l in MDCK cells, and in oocytes that express BGT1, the affinity for GABA is at or below 0.1 mMol/l. At 1 mMol/l substrate, the current induced in BGT 1-expressing oocytes decreased in the following order: betaine > GABA > diaminobutyric acid = β-alanine > proline. At 1 mMol/l betaine, the affinity for sodium was 93.3 mMol/l and the affinity for chloride 76.1 mMol/l (Yamauchi et al. 1992).

In electrophysiological studies, a coupling ratio of Na/Cl/betaine of 3:2:1 was found. Furthermore, these studies revealed significant kinetic differences compared to the neuronal GABA-transporter GAT-1 (Matskevitch et al. 1999).

For the symport process, a transport model of ordered binding was proposed (Matskevitch et al. 1999; Forlani et al. 2001) in which betaine binds first to the extracellular side of a transporter. Sodium binding occurs thereafter, followed by chloride binding. Translocation of betaine and sodium is already observed in the absence of chloride, but chloride augments the translocation. As for other cotransport systems, this model would predict that betaine facilitates sodium and chloride binding to the transporter; corresponding increases in the transporter affinities at higher betaine concentrations have been observed. The exact mechanism of chloride (and sodium and betaine) translocation are unknown to date. It should be mentioned, however, that in squid motor neurons betaine activates a large Cl^- selective current which is sodium-dependent (Petty and Lucero 1999). This observation raises the possibility that the betaine transporter might posses chloride channel properties.

Augmented extracellular osmolality increases betaine uptake into MDCK cells (Nakanishi et al. 1990; Yamauchi et al. 1991; Kempson 1998), rabbit PAP-HT 25 cells (renal papillary epithelium; Ferraris et al. 1996), and porcine chondrocytes (de Angelis et al. 1999) at the level of the transporter by increasing V_{max} but not K_m of betaine uptake (see Fig. 3). The increase in uptake is preceded by an enhanced transcription of BGT1 in MDCK cells followed by increases in BGT1-mRNA. As in the case of the sodium-myo-inositol transporter a "tonicity-responsive enhancer element" (designated TonE) of the gene is involved that spans −69/−50 of the sequence (Burg et al. 1997; Handler and Kwon 2001). In transgenic mice, the 5′-flanking region which contains TonE also mediates osmotic regulation of transcription in vivo (Handler and Kwon 2001). Osmoregulation of BGT1 mRNA has also been observed in renal medulla in the intact rat (Miyai et al. 1996; Moeckel et al. 1997), and in rat hepatic stellate cells (Peters-Regehr et al. 1999). Although the time course and extent of regulation of BGT-1 and SMIT by hypertonicity are quite similar, the regulatory pathways may be different at one point or another (Atta et al. 1999)

In addition to the relatively slow adaptation of betaine transport in vitro, more rapid changes have also been reported. For example, in mouse isolated perfused straight proximal tubules, basolateral betaine uptake (measured as substrate-dependent depolarization) is enhanced when the extracellular osmolality is increased (Völkl and Lang 2001). The speed of regulation suggests nongenomical processes. A similar rapid activation was observed in squid motor neurons (Petty and Lucero 1999). Whether these responses involve protein phosphatases that, potentially by dephosphorylation, could stimulate betaine transport is not clear.

Hypertonic activation of betaine uptake shows some more peculiarities. Under isotonic conditions betaine uptake into MDCK cells proceeds only across the basolateral membrane; after exposure to hypertonicity basal-lateral uptake is increased, but in addition, significant uptake across the apical membrane is observed (Yamauchi et al. 1991). This finding might be related to the existence of various isoforms of BGT1 (see above) that might differ in their cellular location.

Na^+-Myo-inositol symport

Physiological plasma concentrations of myo-inositol range in mammals from 4.5 mMol/l to 6.6 mMol/l, whereas intracellular concentrations up to 133 mMol/l can be found in rat glial (Strange et al. 1991) and renal medullary cells (Bagnasco et al. 1986; Nakanishi et al. 1988; Wirthensohn et al. 1989; Yancey and Burg 1989; Garcia-Perez and Burg 1991; Sizeland et al. 1993; Grunewald et al. 1995; Grunewald and Kinne 1999; Handler and Kwon 2001). This large concentration difference suggests active uptake of myo-inositol into these cells. Similarly, active myo-inositol uptake is observed in rat pancreatic islets, bovine lens epithelial cells, hamster small intestine, rat mesanglial cells, rat hepatocytes, crystalline lense, rabbit peripheral nerve, retinal pigment epithelial cells, rabbit ciliary body, isolated rat Schwann cells, and endothelial cells (for references see Porcellati et al. 1999). This uptake process involves a sodium/myo-inositol symport system (SMIT) which as secondary active transport system employs the transmembrane electrochemical potential difference of sodium for intracellular accumulation of myo-inositol. Apparent K_m values for myo-inositol range from 30 μMol/l in MDCK cells (Kwon et al. 1992) to ~94 μMol/l in rabbit renal brush border (Hammerman et al. 1980) and 104 μMol/l in rabbit TALH cells (Yorek et al. 1999). Interestingly, also D-glucose is a substrate of this transporter (Hammerman et al. 1980; Hager et al. 1995). Phlorizin, the well-known inhibitor of the sodium-D-glucose cotransporter, also inhibits the sodium-myo-inositol transporter (K_i <60 μMol/l; Strange et al. 1991; Hager et al. 1995). Similarities between the two transporters were also detected at the DNA and protein level. There is 46% amino acid identity overall and apparently similar membrane topology of the two transporters; SMIT is a polypeptide of 718 amino acids with a relative molecular mass of 79.5 kDa (Kwon et al. 1992).Tissue abundance of mRNA in dog tissues is kidney medulla > kidney cortex > brain (Kwon et al. 1992). The SMIT cloned recently from bovine lens epithelial cells shows a 92% identity with the MDCK cell transporter (Zhou et al. 1994).

The sodium to myo-inositol stoichiometry of the transporter is 2:1 (Hager et al. 1995). The stoichiometry and electrogenicity of the transporter forms the basis for the extensive intracellular accumulation of myo-inositol. Kinetic and biochemical characteristics vary from tissue to tissue and cell to cell. These variations have recently been explained by the existence of alternate splicing, which generates isoforms that differ in their intracellular protein kinase A and protein kinase C phosphorylation sites at the carboxy terminus (Porcellati et al. 1999). In epithelial cells, the cellular location can also differ; thus, in the renal proximal tubule the Na^+/myo-inositol symport is present in the luminal brush border membrane, whereas in more distal renal cells the basolateral location prevails (see Grunewald and Kinne 1999).

SMIT is increased by augmentation of the extracellular osmolality in MDCK cells (Nakanishi et al. 1989; Yamauchi et al. 1991), glial cells (Strange et al. 1991; Kwon et al. 1992), neuronal cells (Yamashita et al. 1999), mesothelial cells (Matsuoka et al. 1999), bovine lens epithelial cells (Cammarata et al. 2002), and human retinal epithelial cells (Handler and Kwon 2001). In MDCK cells grown on filters, a 25-fold increase in uptake rate was observed within 24 h after exposure of the cells to hypertonic media (Yamauchi et al. 1991).

After restoring isotonicity, the transport rate returns to normal levels within 1 d. In both instances only the V_{max} of the transport is affected and K_m remains constant, suggesting a change in the number of transporters (Nakanishi et al. 1989). In similar studies in rat C6 glioma cells, myo-inositol uptake also increased after exposure of the cells to 440 mOsm/l solutions; however, the time course was slower and maximum uptake was obtained after 48 h (Strange et al. 1991). The increased uptake in response to hypertonicity is preceded by an increased abundance of mRNA for the transporter, which is the direct result of increased transcription of the gene (Yamauchi et al. 1993). Activation of transcription depends on an enhancer element named tonicity responsive enhancer (TonE). The SMIT-gene is regulated by multiple TonEs in its 5′-flanking region (Burg et al. 1997; Zhou and Cammarata 1999; Handler and Kwon 2001). An increase in SMIT m-RNA is also found in a TALH cell line derived from rabbit renal medulla (Yorek et al. 1999) and in mesothelial cells (Matsuoka et al. 1999). In the latter cell line, the first significant increase in mRNA could be observed already 1 h after exposure of the cells to 490 mOsm. The increase in transporter mRNA has also been observed under in vivo conditions, for example, in rat inner medullary collecting duct cells when d DAVP was administered to chronically diuretic rats, a maneuver that rapidly increases the extracellular osmolality in the medulla (Burger-Kentischer et al. 1999). Injection of NaCl into rats increases SMIT-mRNA significantly within 1 h. Comparing the experiments in intact rats with cultured cells, the transporter mRNA seems to respond more rapidly in vivo than in vitro (Yamauchi et al. 1995).

Amino acid transport system A

Amino acid transport system A (System A), which mediates sodium-dependent uptake of neutral amino acids into mammalian cells, appears to be also subject to regulation by extracellular osmolality. In MDCK cells, as well as chondrocytes, an upregulation within 4–6 h of hypertonic exposure was observed, well before any change in BGT1 activity (Chen et al. 1996a; de Angelis et al. 1999; Horio et al. 1997; Kempson 1998). The increase in System A is blocked by inhibitors of RNA and protein synthesis, suggesting that an increase in the number of transporters is part of the mechanism (Kempson 1998). The adaptation of this system appears to be a relatively early response of cells exposed to hypertonicity; therefore, it deserves further investigation.

Regulatory volume decrease

Inorganic osmolytes in RVD

After cell swelling, transport systems are immediately activated that mediate the release of the major intracellular inorganic osmolytes potassium and chloride. Their transmembrane movement occurs either via separate pathways or directly coupled to each other.

K^+ channels

Due to the outwardly directed K^+ gradient in most animal cells, any increase of K^+ channel activity will augment the conductive exit of K^+. In addition, the increase of cell membrane K^+ conductance will hyperpolarize the cell membrane and (even if basal Cl^- conductance is not changing) this hyperpolarization will favor conductive Cl^- efflux. Likewise, if there is an initial increase of Cl^- conductance and if cell Cl^- is above electrochemical equilibrium (as it is in most systems) this will augment Cl^- exit and depolarize the cell membrane. Membrane depolarization in turn will facilitate conductive K^+ efflux. The most effective mechanism of RVD will, of course, be the parallel activation of K^+ and Cl^- channels. Because of the pronounced voltage-mediated coupling between both pathways conductive K^+ and Cl^- release may result in a quasielectroneutral mode of KCl export.

An increase of cell membrane K^+ conductance under hypotonic conditions and its significance for RVD have been reported from a variety of preparations (see Lang et al. 1998a; Lang et al. 1998b for comprehensive reviews). In many studies, these mechanisms were analyzed by means of $^{86}Rb^+$ (or $^{42}K^+$) fluxes, intracellular microelectrode recordings and rapid ion-substitutions, whole-cell patch-clamp measurements, and by use of specific K^+ channel blockers. Here, we will focus on data at the molecular level that were obtained by means of single-channel recordings and by the cloning and expression of some of the volume-activated K^+ channels.

BK_{Ca} or maxi-K^+ channels exhibit large (*big*) unitary conductances in the range of some 100 pS to 250 pS and show under symmetrical high K^+ solutions a linear current-to-voltage relation (see Vergara et al. 1998 for review). In most instances, BK_{Ca} channels are inhibited by Ba^{++}, quinine, and TEA (tetraethylammonium), as well as by (the scorpion peptide) charybdotoxin. They are selectively blocked by (the scorpion toxin) iberiotoxin. BK_{Ca} channels are found in neurons, skeletal muscle, smooth muscle, and in epithelial cells, where they reside in the apical membrane. BK_{Ca} channels are activated by membrane depolarization and micromolar concentrations of cell Ca^{2+}. It is this Ca^{2+} sensitivity that might function as a link to a swelling-induced increase in intracellular calcium. Likewise, the voltage-dependence of BK_{Ca} channels may reflect the coupling mechanism to a hypotonicity-induced activation of Cl^- channels (see above).

BK_{Ca} channels consist of hetero-oligomeric complexes of the pore-forming α-subunits (first cloned from Drosophila as d*Slo* (Atkinson et al. 1991; Adelman et al. 1992) and the auxiliary β-subunits (Garcia-Calvo et al. 1994; Behrens et al. 2002). The α-subunits are expressed in multiple splice variants generating functional diversity among different cells, but they exhibit a high degree of homology among species. Their membrane topology is related to that of voltage-gated K^+ channels (see below; Toro et al. 1998; Vergara et al. 1998; Jensen et al. 2001). They contain seven transmembrane segments (so that the N-terminus is most likely extracellular), a positively charged TM4-segment (that is probably part of the voltage sensor), and four additional hydrophobic segments forming a unique secondary structure at the C-terminal cytosolic tail that appears to mediate Ca^{2+} binding (Schreiber and Salkoff 1997).

The β-subunit is likely to function as an additional modulator of channel characteristics, e.g., with respect to Ca^{2+} sensitivity, but it does not appear to be an obligatory component of all BK_{Ca} channels. Four β-subunits have been cloned that are mainly expressed in smooth muscle, endocrine cells, epithelia, and the central nervous system (Behrens et al. 2002). β-subunits have a proposed topology of spanning the membrane twice, with N- and C-termini inside the cell (Toro et al. 1998; Vergara et al. 1998).

The activation of BK_{Ca} channels under hypotonic conditions could be demonstrated in *Necturus* and rabbit proximal tubule cells (Dubé et al. 1990; Filipovic and Sackin 1991; Kawahara et al. 1991), in principal cells of rabbit and rat cortical collecting tubule (Ling et al. 1992; Schlatter 1993; Stoner and Morley 1995; Hirsch and Schlatter 1997), in clonal kidney (Vero) cells derived from African green monkey (Hafting and Sand 2000), in acinar cells of rat lacrimal gland (Park et al. 1994), in embryonic chick hepatocytes (Pon and Hill 1997), and in guinea pig jejunal villus enterocytes (MacLeod and Hamilton 1999b). In the A_3 cell line derived from rabbit medullary thick ascending limb, hypotonic stress, as well as negative pressure applied to the patch pipette (suction), significantly increased the open probability of BK_{Ca} channels (Taniguchi and Guggino 1989). Comparable effects were observed in intercalated cells of rat cortical collecting tubule (Pácha et al. 1991), in the human colonic cell line CaCo-2 (Bear 1991), and in vascular smooth muscle cells (Kirber et al. 1992).

These results suggest a direct effect of membrane stretch on the channel itself or on a closely associated component as a mode of channel regulation. Also of note is the distinct sensitivity of BK_{Ca} channels to extracellular ATP as it was observed in Vero cells (Hafting and Sand 2000). Thus, ATP release may well function as an autocrine (or paracrine) mechanism of hypotonic channel activation. In addition, a variety of protein kinases and phosphatases appear to have an impact on BK_{Ca} channel activity (Levitan 1994; Vergara et al. 1998; Hafting and Sand 2000) but their precise role in channel regulation remains to be elucidated.

IK_{Ca} channels are activated by cytosolic Ca^{2+} activities in the upper nanomolar range and exhibit an *intermediate conductance* that (at 0 mV) equals 20 pS–80 pS. They are voltage-independent but weakly inwardly rectifying (under symmetrical high K^+ conditions), (see Vergara et al. 1998; Jensen et al. 2001 for review). The apparently large range of unitary conductances reported for these channels is in part due to differences in experimental design and the very pronounced dependence of these channels on the actual (extracellular) K^+ activity. IK_{Ca} channels are weakly inhibited by quinine but efficiently blocked by charybdotoxin and (more selectively) by clotrimazole. The latter compound was found to have a high affinity to the "Gardós channel" (this channel is the IK_{Ca} of erythrocytes and actually the first Ca^{2+}-dependent K^+ transport described; Gardos 2002) and was therefore considered to be of potential use for the treatment of sickle cell anemia (Jensen et al. 2001). IK_{Ca} channels are virtually insensitive to the bee venom toxin apamin, which discriminates them from most isoforms of SK_{Ca} channels (see below; Jensen et al. 2001). Interestingly, IK_{Ca} channels are selectively activated by the channel modulator 1-EBIO (1-ethyl-benzimidazolinone; Jensen et al. 2001).

The first IK_{Ca} channel was cloned from human tissues (*KCNN4*; Ishii et al. 1997; Joiner et al. 1997; Logsdon et al. 1997; Jensen et al. 1998) and shortly thereafter the orthologues of mouse and rat followed (Warth et al. 1999; Vandorpe et al. 2002). The human IK_{Ca} channel is 88% and 90% identical to the mouse and rat channel, respectively, and 40% to 42% identical to SK_{Ca} channels. The short cytosolic N-terminus is followed by six transmembrane segments and a long intracellular C-terminus (Vergara et al. 1998; Jensen et al. 2001). The channel pore is most probably located in a hydrophobic pocket between TM5 and TM6 and the channel is likely to function as a homotetramer. The Ca^{2+} sensitivity of IK_{Ca} channels appears to be mediated by calmodulin binding to a C-terminal domain. Using Northern blot analyses, IK_{Ca} channels were mainly detected in tissues rich in epithelia and endothelia (Jensen et al. 2001). This is of note because of the known high rates of vectorial transport in such systems and the resultant need for an effective osmotic cell homoiostasis.

There are several reports which identify this type of channel as the mediator of conductive hypotonicity-induced K^+ release. In the human epithelial cell line Intestine 407, hypotonic swelling led to a significant increase of cell Ca^{2+}, thereby activating IK_{Ca} channels (Okada et al. 2001). Human T lymphocytes express a Ca^{2+}-activated K^+ conductance that was activated by hypotonic conditions (Khanna et al. 1999). The K^+ conductance was slightly inwardly rectifying and blocked by charybdotoxin as well as clotrimazole. In transformed Madin-Darby canine kidney (MDCK-F) cells, hypotonic stress increased cell Ca^{2+} and activated a 53 pS K^+ channel that was inwardly rectifying (Schwab et al. 1993) and blocked by charybdotoxin, Ba^{2+}, and TEA (Schwab and Oberleithner 1996; Schwab et al. 1999; Schneider et al. 2000). Because MDCK-F cells employ cell volume regulatory mechanisms for locomotion, cell migration could be shown to be inhibited with an identical pharmacological profile (Schwab and Oberleithner 1996). Moreover, the channel appeared to be activated by 1-EBIO (Schwab et al. 1999). In the distal nephron cell line A6 derived from *Xenopus* kidney, cell swelling activated a K^+ channel that (in symmetrical 130 mMol/l K^+) was slightly inwardly rectifying with a unitary conductance of 29 pS (at

0 mV) and inhibited by quinine (Nilius et al. 1995). It is tempting to speculate that the channel may be Ca^{2+}-sensitive because, in A6 cells, hypotonic stress was shown to elicit marked increases of cell Ca^{2+} that paralleled RVD and the activation of a TEA-sensitive K^+ conductance (Yu and Sokabe 1997; Urbach et al. 1999).

In some systems, the putative role of IK_{Ca} channels in the RVD process is not yet conclusively defined. In MDCK cells for instance, hypotonic stress was reported to transiently activate K^+ channels that were Ca^{2+}-sensitive and in cell-attached patches (with 150 mMol/l K^+ in the pipette); these channels were inwardly rectifying with a unitary conductance of 29 pS at 0 mV (Weiss and Lang 1992). From the same system, Banderali and Roy (Banderali and Roy 1992a) reported a Ca^{2+}-sensitive channel that showed a significant increase of open probability upon hypotonic cell swelling as well and that was blocked by quinine (Roy and Banderali 1994). In symmetrical 145 mMol/l K^+, however, the channel clearly exhibited a linear current-to-voltage relation with a unitary conductance of 24 pS (Banderali and Roy 1992a; Roy and Banderali 1994). In an early report on Ehrlich ascites tumor cells, RVD was found to involve the activation of an inwardly rectifying K^+ channel that (under symmetrical high K^+ conditions) exhibited unitary conductances of 40 pS and 15 pS in the negative and positive voltage range, respectively. The channel appeared to be Ca^{2+}-activated and its open probability did not depend on membrane voltage (Christensen and Hoffmann 1992). In a more recent study on the same preparation, however, Ca^{2+} signaling did not seem to play any significant role in RVD at all, and hypotonicity-induced K^+ release was found to be inhibited by Ba^{2+} but clearly not by charybdotoxin or clotrimazole (Jorgensen et al. 1997). In some other systems, cell swelling activates K^+ channels of intermediate conductance that apparently are not related to IK_{Ca}. In *Necturus* proximal tubule for instance, cell-attached patches on the basolateral membrane revealed the presence of a K^+ channel that could be activated by negative pressure as well as by a reduction in bath osmolarity (Sackin 1989). There was no detectable voltage dependence of channel gating and the channel did not appear to be Ca^{2+}-sensitive. Under symmetrical high K^+ conditions (70 mMol/l), the channel exhibited a linear current-to-voltage relation and a conductance of 36 pS (Filipovic and Sackin 1992). In cell-attached patches on *Xenopus* proximal tubule, a basolateral K^+ channel was observed that was activated by negative pressure and blocked by Ba^{2+} and was supposed to be involved in RVD (Kawahara 1990). With 100 mMol/l K^+ in the pipette, the channel showed little if any inward rectification with a unitary conductance of 28 pS and it did not appear to be Ca^{2+}-sensitive.

SK_{Ca} channels under symmetrical high K^+ conditions exhibit a *small conductance* of some 4 pS–18 pS (at 0 mV) and are activated by nanomolar concentrations of Ca^{2+}. They are inwardly rectifying and voltage-independent (see Vergara et al. 1998 for review). In electrically excitable cells, SK_{Ca} channels mediate the slow after-hyperpolarization following action potentials. SK_{Ca} channels have been recently cloned from—or identified in—rat brain and colon (rSK1–3 and rSK4, respectively), as well as human brain (hSK1) and placenta (hSK4; Köhler et al. 1996; Warth et al. 1999; Okada 1997; Joiner et al. 1997). The topology of SK_{Ca} channels is similar to that of voltage-gated K^+ channels with six transmembrane helices, an intracellular C- and N-terminus, positively charged residues in the TM4 segment, and the pore-forming region between TM5 and TM6. Primary sequences, however, are significantly different. The channels are likely to function as homotetramers, and gating of SK_{Ca} channels involves an interaction of calmodulin with C-terminal domains. SK_{Ca} isoforms differ markedly with respect to their sensitivity to apamin: SK2 is highly sensitive to the bee venom toxin (with a K_d of 60 pMol/l for rSK2) whereas SK1 and SK4 are not significantly inhibited in concentrations up to 100 nMol/l; SK3 exhibits an intermediate sensitivity (Vergara et al. 1998). The sensitivity to apamin is defined by a distinct group of amino acids in the deep pore of the channels, which also determine channel sensitivity to D-tubocurarine, an additional, quite selective blocker. In a recent study, a full-length 2.1 k-bp cDNA (hSK2) highly homologous to rat brain rSK2 was isolated from a human liver cDNA library and identical cDNAs were obtained from human primary hepatocytes, human HuH-7 hepatoma cells, and human Mz-ChA-1 cholangiocarcinoma cells (Roman et al. 2002). Stable transfection of CHO cells with hSK2 resulted in the expression of an apamin-sensitive K^+ conductance as revealed by whole-cell patch-clamp recordings. Thus far, there is only limited evidence for the involvement of SK_{Ca} in volume regulation. Although in Mz-ChA-1 cells hypotonic stress led to a prominent increase of K^+ conductance, and this effect as well as RVD was partially inhibited by apamin (at 50 nMol/l and 100 nMol/l, respectively), Ba^{2+} (at 5 mMol/l) exhibited a complete block of RVD, suggesting that additional K^+ channels contribute to this process (Roman et al. 2002).

Two members of *voltage-gated* K^+ channels could be defined as mediators of RVD, namely Kv1.3 and Kv1.5. They belong to the group of delayed rectifier channels that were first cloned from (electrically) excitable tissues (Stühmer et al. 1989; Tamkun et al. 1991) where they stabilize resting potential and mediate the rapid repolarization of action potentials. K^+ channels of the Kv group typically exhibit six transmembrane helices, an intracellular C- and N-terminus, and a positively charged TM4 segment that most likely is part of the gating machinery. Kv channels were found to be also expressed in electrically nonexcitable cells such as T-lymphocytes (Lewis and Cahalan 1995). There is increasing evidence suggesting that they play a role in cell volume regulation (Felipe et al. 1993; Lewis and Cahalan 1995). It was found, for instance, that a mouse T-lymphocyte cell-line (CTLL-2) that does not express voltage-dependent K^+ channels did not exhibit any significant RVD under hypotonic conditions. Transient transfection of these cells with Kv1.3, however, reconstituted their capability of an almost complete volume regulatory response and led to the generation of significant voltage-activated K^+ currents (Deutsch and Chen 1993). Moreover, the RVD was blunted in the presence of 50 nMol/l charybdotoxin. In a study on human T-lymphocytes both hSK2 (see above) and Kv1.3 were found to contribute to K^+ conductance and RVD (Khanna et al. 1999). In this investigation, the contribution of Kv1.3 was quantified in a differential approach by use of the specific blocker (scorpion toxin) margatoxin (applied at 5 nMol/l). Stable transfection of a mouse fibroblast cell line (Ltk^-) with Kv1.5 prevented isotonic cell swelling (that was elicited by the application of dexamethasone) and elicited sizeable voltage-activated K^+ currents (Felipe et al. 1993). Both effects of channel expression were blocked by 60 μMol/l quinine.

The I_{sK} protein was first cloned from rat kidney by use of the *Xenopus* oocyte expression system (Takumi et al. 1989). Its most remarkable feature is its *small size* of 126–130 amino acids (depending on the species) with a single transmembrane α–helical domain (Busch and Maylie 1993; Busch and Suessbrich 1997). In various systems, expression of I_{sK} elicited a K^+ conductance upon membrane depolarization that was very slowly activating with time constants of 10 s and more (Tai et al. 1997). I_{sK} functions as a regulatory subunit of K_vLQT1 (which exhibits the typical topology of voltage-dependent K^+ channels and a tissue distribution that is very similar to that of I_{sK}) thus forming the functional "minK" channels (Barhanin et al. 1996; Tai et al. 1997). Alternative partners for both K_vLQT1, as well as I_{sK}, however, may also exist.

In *Xenopus* oocytes that were injected with I_{sK} cRNA (and that were later found to endogenously express a K_vLQT1 analog; Sanguinetti et al. 1996), slowly developing voltage-activated K^+ currents were detected (Busch et al. 1992). Hypotonic stress led to a pronounced increase of these currents, to an accelerated activation, and to a shift in voltage dependence to more negative membrane voltages. The above effects were completely abolished in Ca^{2+}-free solutions (Busch et al. 1992). A significant contribution of minK channels to the RVD process was also reported from vestibular dark cells of gerbil inner ear (Shiga and Wangemann 1995; Wangemann et al. 1995) and from mouse tracheal epithelium (Lock and Valverde 2000). In the latter system, RVD was insensitive to Ba^{2+} and apamin and only weakly inhibited by TEA, which is to be expected for I_{sK}. In contrast, clofilium (a quaternary ammonium blocker, used at 100 μMol/l), which is a rather selective blocker of the K_vLQT/I_{sK} complex, potently inhibited RVD. Moreover, in tracheal epithelial cells, the I_{sK} (–/–) knockout mouse RVD was no more detectable (Lock and Valverde 2000).

In a recent study, TASK-2 (KCNK5) was identified as a mediator of hypotonicity-induced conductive K^+ release in Ehrlich ascites tumor cells (Niemeyer et al. 2001). TASK, for TWIK (*T*andem of P domains in *W*eak *I*nward rectifier *K*$^+$ channels)-related acid-sensitive K^+ channels, belong to the group of two pore-domain K^+ channels (K_{2P}; Lesage and Lazdunski 2000). These channels have four transmembrane segments and they operate as dimers. There is a widespread distribution of K_{2P} channels in both (electrically) excitable and nonexcitable tissues. K_{2P} channels appear to define the background K^+ conductive properties of many cell membranes. They are insensitive to most classical K^+ channel blockers (Lesage and Lazdunski 2000). When expressed in HEK293 cells, TASK-2 elicited hypotonicity-induced currents that are very similar to those in Ehrlich cells, e.g., with respect to ion sensitivity ($K^+ > Rb^+ >> Cs^+ > NH_4^+ > Na^+ \cong Li^+$) and inhibition by clofilium, with an IC_{50} of 25 μMol/l. In addition, mTASK-2 (but not mTASK-1 and mTASK-3) could actually be identified in Ehrlich cells (Niemeyer et al. 2001).

In conclusion, hypotonicity-induced K^+ channels represent an effective mechanism for the selective release of K^+ during RVD. They differ considerably with respect to unitary conductance, gating mechanism, and molecular organization. On the other hand, the high degree of variability among the numerous proteins actually recruited for conductive K^+ release underlines the general importance of this mechanism for RVD (Wehner 1998).

Cl^- channels

The apparently most abundant form of hypotonicity-induced Cl^- channels belong to a group that has been named VSOR (volume-sensitive outwardly-rectifying) Cl^- channels (Okada 1997), VRAC (for "volume-regulated anion channels" (Nilius and Droogmans 2001), or VSOAC (volume-sensitive organic osmolyte-anion channels; Jackson et al. 1994). These channels are referred to as mediators of $I_{Cl,swell}$ (Nilius et al. 1999) or $I_{Cl,vol}$ (Hoffmann and Mills 1999). They appear to be ubiquitously expressed (see Strange et al. 1996; Okada 1997 for comprehensive reviews; see also Nilius and Droogmans 2001). An example for such a channel is given in Fig. 4.

Their significant outward rectification persists under symmetrical Cl^- conditions. An interesting feature of VSOR Cl^- channels is their pronounced inactivation when membrane voltage is stepped from very negative values to above +50 mV. The channels exhibit intermediate conductances in the range of 10 pS to 40 pS at negative and 40 pS to 80 pS at

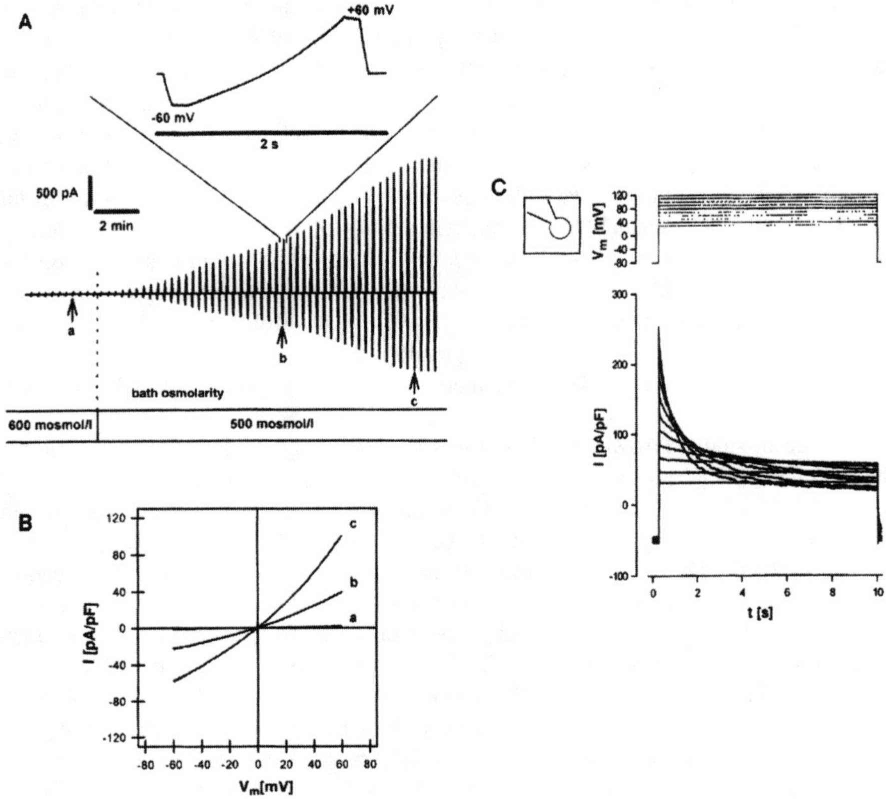

Fig. 4 A Activation of anion conductance in rat inner medullary collecting duct (IMCD) cells under hypotonic conditions. Whole-cell patch-clamp recording with symmetrical CsCl solutions; the osmolarity of the pipette solution was 600 mOsm/l. At the time indicated, osmolarity of the superfusate was reduced from 600 mOsm/l to 500 mOsm/l. Voltage ramps from −60 to +60 mV and 2 s duration were applied every 15 s. **B** Current-voltage relationship before (*line a*), at 6.5 min (*line b*), and at 11 min (*line c*) of hypotonicity; currents were referred to membrane capacitance. **C** Time-dependent inactivation of hypotonicity-induced Cl^- conductance at depolarizing voltages. Membrane voltage was stepped from −80 mV to positive values in the range of +30 to +120 mV with 10 mV increment after each pulse. Pulse duration and interval between pulses were 10 s and 15 s, respectively (**A** and **B** from Boese et al. 1996b; **C** from Boese et al. 1996a)

positive membrane voltages. The activation of VSOR Cl^- channels strictly depends on the nonhydrolytic binding of ATP at physiological intracellular concentrations (Jackson et al. 1994; Jackson et al. 1996; Meyer and Korbmacher 1996). Extracellular ATP in contrast, induces an open-channel block and it does so in its Mg^{2+}-free form (Nilius et al. 1994a; Jackson and Strange 1995a; Jackson et al. 1996; Tsumura et al. 1996). VSOR Cl^- channels exhibit the permeability sequence $SCN^- > I^- > NO_3^- > Br^- > Cl^- > HCO_3^- > F^-$ > isethionate > gluconate > glycine > taurine > aspartate, glutamate (Boese et al. 1996a; Strange et al. 1996; Okada 1997; Nilius and Droogmans 2001). The halide selectivity corresponds to Eisenman's sequence I, suggesting a relatively weak interaction of permeant ions with cationic sites inside the channel when compared to their dehydration energy (Wright and Diamond 1977). The physiological significance of the permeability of the channels to certain amino acids that function as organic osmolytes will be discussed below (see the section entitled "Swelling activated release").

VSOR Cl^- channels are inhibited by submillimolar concentrations of SITS and DIDS. This inhibition is reversible and it is voltage-dependent so that outward currents are more sensitive than inward currents (which is supposed to reflect the preferred entrance of these compounds into the channel mouth at positive membrane voltages; Boese et al. 1996b; Okada 1997). In a voltage-independent fashion, the channel is also efficiently inhibited by NPPB and, in some preparations, by DPC and its derivatives flufenamate and niflumate. Interestingly in some systems, VSOR Cl^- channels were found to be blocked by inhibitors of p-glycoprotein (the product of the MDR1 gene), like tamoxifen, 1,9-dideoxy-forskolin, verapamil, and quinidine, which led to the hypothesis of a close molecular correlation between these membrane proteins. This notion appeared initially to be strongly supported by experiments in which p-glycoprotein was heterologously expressed in various cell systems (Gill et al. 1992; Valverde et al. 1992). Currently even a possible role of p-glycoprotein as a regulator of VSOR Cl^- channels is questioned (Okada 1997; Nilius and Droogmans 2001).

Despite the rather ubiquitous presence of VSOR Cl^- channels, their molecular entity still remains a mystery. One of the reasons for this is the absence of any specific inhibitor of VSOR Cl^- channels. A possible molecular candidate is I_{Cln}, a protein that was originally cloned from a Madin Darby canine kidney (MDCK) cell library and that elicited a nucleotide-sensitive ("n"), outwardly rectifying chloride current (with rapid inactivation at positive voltages) when expressed in *Xenopus* oocytes (Paulmichl et al. 1992). Because of the presence of VSOR Cl^- channels (or functionally related channels) in almost all cells, however, it was always difficult to decide whether the I_{Cln} protein reflects a channel regulator or the channel itself (Strange 1998; Fürst et al. 2000b). A possible role of I_{Cln} as a regulator has been inferred from the observation that, in most instances, it appears to be localized in the cytoplasm rather than in the plasma membrane (Krapivinsky et al. 1994; Buyse et al. 1997). On the other hand, a hypertonicity-induced translocation of the I_{Cln} protein from the cytosol to the plasma membrane has also been reported (Laich et al. 1997; Musch et al. 1997; Musch et al. 1998). This observation per se could, of course, equally well reflect the insertion of a channel or the movement of a regulatory factor. Hydrophobicity analysis indicated that I_{Cln} lacks transmembrane helices that are believed to form the pore of most vertebrate ion channels (Paulmichl et al. 1992). Accordingly, a putative model of an I_{Cln} channel was proposed consisting of a homodimer with each monomer containing four β-strands so that an eight-stranded, antiparallel β-barrel transmembrane pore is formed (Paulmichl et al. 1992; Fürst et al. 2000b). Interestingly, when purified I_{Cln} was reconstituted in lipid bilayers, single channel events could be detected that exhibited a significant rectification; ion substitutions, however, revealed that the channels observed were rather cation- than anion-selective (Li et al. 1998; Fürst et al. 2000a). In heart lipid bilayers, however, chloride selectivity was found (Garavaglia et al. 2002).

ClC-2 and ClC-3 were also considered as molecular candidates for VSOR Cl^- channels. The first member of this group of voltage-gated channels (ClC-0) was cloned from the electric ray *Torpedo* (Jentsch et al. 1990). Typically, ClCs exhibit ten to twelve transmembrane helices and the N- and C-termini are located in the cytosol. Most ClCs (except some prokaryotic isoforms) contain two hydrophobic domains close to the C-terminus that may be involved in their cellular sorting (Schmidt-Rose and Jentsch 1997; Jentsch et al. 1999). ClCs were supposed to function as dimers, with each monomer showing a separate pore.

This structure could recently be proven by X-ray analysis on two bacterial ClC isoforms with 3 Å resolution (Dutzler et al. 2002). ClC-2 is a broadly expressed Cl^- channel (Jentsch et al. 1999) that is activated by strong (nonphysiological) hyperpolarizations, acidic extracellular pH, and cell swelling (Gründer et al. 1992). However, it clearly differs from VSOR Cl^- channels with respect to its voltage dependence (it is *inwardly* rectifying) and its anion selectivity ($P_{Cl} > P_I$). An actual role of ClC-2 in cell volume regulation was a matter of debate (Jentsch et al. 1999), but recent evidence supports this notion. In *Xenopus* oocytes heterologously expressing ClC-2, for instance, hypotonic stress elicited a significant increase of cell membrane Cl^- conductance (Stegen et al. 2000). Likewise in Sf9 cells, expression of ClC-2 led to swelling-activated chloride currents that could be specifically inhibited by anti-ClC-2 antibodies (Xiong et al. 1999). In mouse atrial and ventricular myocytes, a Cl^- current was observed that was activated by hyperpolarization of membrane voltage and current activation was increased by cell swelling (and inhibited by cell shrinkage). The selectivity sequence of currents was $Cl^- > I^- >>$ aspartate, and RT-PCR and Northern blot analysis confirmed transcriptional expression of ClC-2 in both tissues (Duan et al. 2000). In rat HTC hepatoma cells, hypotonic stress elicited inwardly rectifying Cl^- currents that could be inhibited by anti-ClC-2 antibodies. Moreover, a cDNA closely homologous to rat brain ClC-2 was isolated and ClC-2 mRNA and membrane protein expression was demonstrated by in situ hybridization, immuno-cytochemistry, and Western blot analysis (Roman et al. 2001).

ClC-3 exhibits many similarities to VSOR Cl^- channels including outward rectification, unitary conductance, anion selectivity, voltage-dependent inactivation, and inhibition by extracellular nucleotides, stilbene derivatives, and tamoxifen (Strange 1998; Jentsch et al. 1999; Hume et al. 2000). There are some significant differences remaining, however. Among these is the enormous basal activity of ClC-3 when compared to VSOR Cl^- channels, the latter being undetectable under isotonic conditions. Moreover, ClC-3 appears to be inhibited by PKC, whereas PKC stimulates VSOR Cl^- channels in many preparations (Okada et al. 1998; Okada 1998; Strange 1998).

In most systems, hypotonic stress elicits a significant increase in cell Ca^{2+}. This in turn may lead to the activation of Ca^{2+}-dependent Cl^- channels (CaCCs) that may contribute significantly to the overall volume response (Jentsch 1996; Hoffmann and Mills 1999; Nilius and Droogmans 2001). In Ehrlich ascites tumor cells and in the mouse inner medullary collecting duct cell line mIMCD-K2, for instance, the contribution of both VSOR Cl^- channels and CaCCs to the hypotonicity-induced Cl^- release could be determined in a quantitative fashion (Pedersen et al. 1998; Boese et al. 2000). CaCCs activate slowly at positive and inactivate rapidly at negative voltages. They exhibit strong outward rectification with unitary conductances in the range of 3 pS to 10 pS and the ion permeability sequence is $I^- > Cl^- >$ gluconate (see Nilius and Droogmans 2001 for review). Channel activity is high at positive and low at negative voltages. CaCC currents are blocked by DIDS, NPPB, niflumate, and tamoxifen. The molecular correlate of CaCCs have not yet been unequivocally identified. The first member of a gene family of Ca^{2+} dependent Cl^- channels was cloned from bovine trachea (Cunningham et al. 1995) and since then a number of isoforms followed (Fuller and Benos 2000; Fuller et al. 2001). These "CLCA" are heterodimers of some 90 kDa and 35 kDa that are cleaved from a common precursor (at the inner loop between TM3 and TM4) and the most likely topology are five transmembrane regions with an extracellular N- and an intracellular C-terminus. There are a number of similarities between CLCAs and CaCCs but also significant differences, including unitary conductances which appear to be higher in CLCAs. An actual role of CLCAs in cell volume regulation remains to be demonstrated (Nilius and Droogmans 2001; Papassotiriou et al. 2001).

BCl (or maxi Cl^-) channels that are activated under hypotonic conditions have been reported from cardiac myocytes of new-born rats (Coulombe and Coraboeuf 1992), mouse N1E115 neuroblastoma cells (Falke and Misler 1989), rat cortical astrocytes in primary culture (Jalonen 1993), the rabbit renal cortical collecting duct cell line RCCT-28A (Schwiebert et al. 1994), and primary cultures of bovine pigmented ciliary epithelium (Zhang and Jacob 1997). Typically, BCl channels exhibit unitary conductances in the range of 100 pS to 400 pS and a linear current-to-voltage relationship. In most instances, up to five identical subconductance levels were detected. BCl channels have a bell-shaped voltage dependence with very low activity outside the range of –40 mV to +40 mV. The anion-to-cation selectivity is higher than ten and the channels are inhibited by DIDS, NPPB, and DPC, but apparently not by tamoxifen. BCl channels may be activated by membrane stretch or by patch excision from nonswollen cells. Based on the effects of cytochalasins, a direct regulatory interaction with the actin cytoskeleton is assumed (Strange et al. 1996; Nilius et al. 1997). BCl channels may be related to the voltage-dependent anion channel (VDAC), a porin-like channel found in eukaryotic mitochondria or to the brain-derived voltage-dependent anion channel 1 (BR1-VDAC), which is also expressed in lymphocytes, kidney, heart, and skeletal muscle (Dermietzel et al. 1994; Thinnes et al. 2001).

Hypotonicity-induced SCl (or mini Cl^-) channels with unitary conductances in the range of 2 pS to 10 pS have been reported from single-channel patch-clamp studies on the ventricular membrane of choroid plexus (Christensen et al. 1989), Ehrlich ascites tumor cells (Christensen and Hoffmann 1992), and pigmented and nonpigmented bovine ciliary epithelium (Zhang and Jacob 1997). It is because of these low conductances that the biophysical properties and the actual contribution of SCl channels to macroscopic conductances were sometimes hard to define. In the human colon cell-line HT_{29}, for instance, the Cl^- channels activated by cell swelling appeared to be too small to be detected by use of single-channel patch-clamp techniques (Kubitz et al. 1992). Stationary noise-analysis of whole-cell currents was considered a way out and unitary currents in the range of 0.26 pS to 6 pS were determined by this method in mouse and human T and B lymphocytes (Lewis et al. 1993; Schumacher et al. 1995), human neutrophils (Stoddard et al. 1993), bovine chromaffin cells (Doroshenko and Neher 1992), and various endothelial, epithelial, fibroblast, and blood cells (Nilius et al. 1994b).

As was outlined in detail by Strange and coworkers (Jackson and Strange 1995b; Jackson et al. 1996), however, this approach may lead to a marked underestimation of unitary conductances (and to a considerable overestimation of the actual number of channel molecules involved) because some basic assumptions made—the number of channels remains constant and these channels respond to cell swelling with a gradual increase in open probability—may not be valid (see also Nilius et al. 1997). Instead, nonstationary noise-analysis may be more appropriate as could be shown for (at least some) VSOR Cl^- channels which respond to hypotonic stress with an abrupt shift of open probability from virtually zero to a value very close to unity so that the number of channels contributing to the overall membrane response is increasing, i.e., it is not constant (Jackson and Strange 1995b; Boese et al. 1996a). Accordingly, some of the above data obtained by noise-analysis may need reevaluation, and the apparently low unitary conductances determined may actually be considerably higher.

Taken together, despite the enormous efforts undertaken in recent years to define the biophysical and molecular properties of hypotonicity-induced Cl^- channels, the overall image is still elusive and, in some instances, rather conflicting results were obtained. One of the most important steps to be achieved will be towards the molecular understanding of VSOR Cl^- channels, which appear to be the most commonly used mechanism of conductive Cl^- release during RVD.

K^+-Cl^- symport

Electroneutral KCl release appears to be one of the major features during RVD in many systems (Lang et al. 1998a; Lauf and Adragna 2000; Hoffmann and Mills 1999; Lang et al. 1998b).

K^+-Cl^- symport was first defined in erythrocytes as Cl^- dependent and ouabain-insensitive K^+ transport and, thereafter, detected in many other cell types (see Lauf and Adragna 2000 for review). Four isoforms of K^+-Cl^- symporters (KCC1–KCC4) have been cloned to date which exhibit significant molecular similarities to the Na^+-Cl^- symporter (NCC) and to the known mediators of Na^+-K^+-$2Cl^-$ symport (NKCC1 and NKCC2; see the section entitled "Na^+-K^+-$2Cl^-$ symport"). All together, they form the superfamily of cation-Cl^- cotransporters (CCCs) that appears to originate from an ancient short transport protein found in cyanobacteria (Gillen et al. 1996). CCCs contain twelve putative transmembrane domains with large N- and C-terminal regions in the cytosol. Typically, however, KCCs have a large extracellular loop between TM5 and TM6 that is variably N-glycosylated between the different isoforms (Gillen et al. 1996; Lauf and Adragna 2000). TM2 probably reflects the binding/transport site for K^+, whereas TM4 and TM7 are likely to be involved in Cl^- transport (Lauf and Adragna 2000). KCC1 appears to be ubiquitously expressed (Haas and Forbush III 2000; Lauf and Adragna 2000), while KCC2 was found exclusively in the brain (Payne 1997; Jarolimek et al. 1999). KCC3 was most prominent in heart and kidney, but also detectable in skeletal muscle, placenta, liver, lung, pancreas, endothelial cells, and in the central nervous system (Hiki et al. 1999; Mount et al. 1999; Pearson et al. 2001). KCC4 exhibited a rather restricted expression pattern with significant amounts of transcript only in muscle, lung, heart, and kidney (Mount et al. 1999). The likely function of KCC2 in the central nervous system is to decrease cell Cl^- below its electrochemical equilibrium and thus to facilitate the hyperpolarizing potency of GABA or glycine-gated Cl^- channels.

KCC1 (Gillen et al. 1996; Holtzman et al. 1998; Mount et al. 1999; Su et al. 1999; Mercado et al. 2000; Lauf et al. 2001) and KCC3 (Mount et al. 1999; Race et al. 1999), as well as KCC4 (Mercado et al. 2000), are clearly activated under hypotonic conditions (when heterologously expressed in HEK-293 cells or *Xenopus* oocytes). KCC2 was found to be insensitive to cell swelling (Payne 1997). With respect to the regulation of KCCs, *dephosphorylation* by specific phosphatases (e.g., PP1A) appears to be the trigger for stimulation (Lauf and Adragna 2000). This is an intriguing difference to NKCCs which are activated by *phosphorylation* and which are mediators of the opposite cellular volume response, namely RVI. Also certain protein kinases do participate in KCC regulation. In rat vascular smooth muscle cells, it was found that protein kinase G (PKG)-1 posttranscriptionally led to an acute upregulation of KCC3 mRNA (Di Fulvio et al. 2001b). Moreover, the NO/

cGMP signaling pathway participates in the mRNA regulation of KCC1, which is the major mediator of RVD in these cells (Di Fulvio et al. 2001a).

Cation channels

In some systems, hypotonic stress leads to the activation of nonselective cation channels. At first sight this may be surprising because, due to the steeper electrochemical Na^+ than K^+ gradient across the cell membrane, nonselective cation channels are expected to mediate a net uptake rather than a release of osmolytes. It has to be considered, however, that some of these channels exhibit a slightly higher permeability for K^+ when compared to Na^+. Moreover, certain nonselective cation channels are permeable to divalent cations so that they may play a role in Ca^{2+} signaling. Some nonselective cation channels are sensitive to membrane stretch.

A cation channel with a P_K/P_{Na} of 4.7 was reported from cultured rat mesangial cells (Craelius et al. 1993). In cell-attached patches, this channel could be activated by hypotonic stress and by negative pressure (suction) applied to the patch pipette. The channel exhibited unitary conductances of 76 pS and 40 pS with high K^+ and high Na^+ in the pipette, respectively. It was impermeable to Ca^{2+} and Ba^{2+} and its mean open time increased with membrane depolarization. In the rat pheochromocytoma cell-line PC12, cell swelling, as well as negative pressure, activated a cation channel of 46 pS with high K^+ and 27 pS with high Na^+ in the (cell-attached) pipette (Cornet et al. 1993). In ventricular myocytes of the rabbit, hypotonicity induced a cation conductance with a P_K/P_{Na} of 5.9 that was blocked by 10 μMol/l Gd^{3+} (Clemo and Baumgarten 1997).

A second group of cation channels does not discriminate significantly among small monovalent cations, but these channels exhibit a sizeable permeability to Ca^{2+} (and/or Ba^{2+}). Channels of this type were found in Ehrlich ascites tumor cells (Christensen and Hoffmann 1992), rat hepatocytes (Bear 1990), single proximal tubule cells of frog kidney (Robson and Hunter 1994), porcine cerebral capillaries (Popp et al. 1992; Popp et al. 1993), rat atrial cells (Kim and Fu 1993; Kim 1993), guinea-pig gastric smooth muscle cells (Yamamoto and Suzuki 1996), mouse N1E115 neuroblastoma cells (Falke and Misler 1989), as well as in *Rana* and *Xenopus* oocytes (Taglietti and Toselli 1988; Yang and Sachs 1989; Hurst and Hunter 1990; Schütt and Sackin 1997). Unitary conductances were in the range of 15 pS to 40 pS and, in most instances, channels were blocked by micromolar concentrations of Gd^{3+}.

Recently, first molecular correlates to hypotonicity-induced, nonselective, and Ca^{2+}-permeable cation channels could be identified in human, rat, mouse, and chicken. These are the "osmosensitive, transient receptor potential channel 4" (OTRPC4; Strotmann et al. 2000), TRP 12 (Wissenbach et al. 2000), and the "vanilloid receptor-related, osmotically activated channel" (VR-OAC; Liedtke et al. 2000). They belong to a family of gene products that are structurally related to the light-activated transient receptor potential (Trp) cation channel from *Drosophila melanogaster* (Montell and Rubin 1989). Typically, these osmo-sensitive channels have six putative membrane-spanning domains, a cytosolic N- and C-terminus, and a pore loop between TM5 and TM6; of note, there are three ankyrin-repeat domains close to the N-terminus (Liedtke et al. 2000). The channels appear to be abundantly expressed in kidney but also in lung, liver, spleen, fat tissue, heart, brain, testis, and endothelial cells. When expressed in HEK293 cells, OTRPC4 channels were activated under hypotonic conditions and exhibited outward rectification with unitary conductances

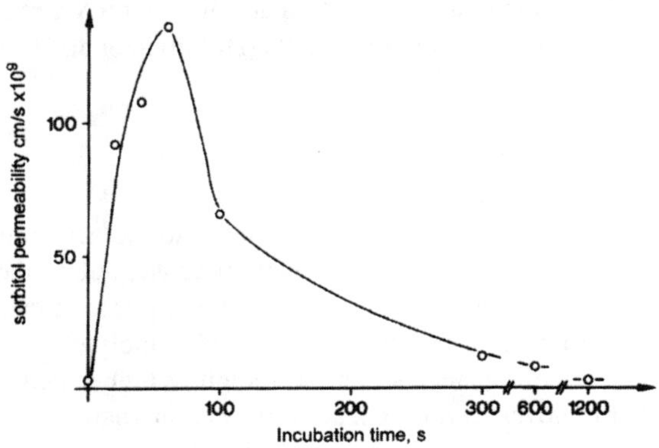

Fig. 5 Time course of activation of sorbitol efflux from IMCD cells during exposure to hypotonic solution. IMCD cells isolated at 600 mOsm/l were exposed at time 0 to a medium of 300 mOsm/l. Membrane permeability was calculated based on the difference between intracellular and extracellular concentration of sorbitol (from Czekay et al. 1994)

of 30 pS for inward and 90 pS for outward currents (Strotmann et al. 2000). P_{Ca}/P_{Na} was 0.8 and $P_K : P_{Cs} : P_{Na} : P_{Li}$ equaled 1.9:1.3:1:0:0.9 (Strotmann et al. 2000; Nilius et al. 2001). The cation channel was insensitive to membrane stretch. It was inhibited by Gd^{3+} (70%) and La^{3+} (30 %) at 100 μMol/l each and completely blocked by 10 μMol/l of the poly-cation ruthenium red, an effective inhibitor of the vanilloid receptor VR1 and the vanilloid receptor-like protein VRL-1 (Strotmann et al. 2000). Expression of VR-OAC in CHO cells led to hypotonicity-induced whole-cell currents with strong outward rectification (and with dual rectification in the absence of extracellular Ca^{2+}; Liedtke et al. 2000). In cell-attached measurements (at +80 mV), unitary current events corresponding to a conductance of 310 pS became detectable. VR-OAC channels were blocked by 500 μMol/l Gd^{3+}. Transient transfection of CHO and HEK293 cells with TRP 12 led to increased cytosolic Ca^{2+} levels when compared to nontransfected cells and, upon hypotonic stimulation, cell Ca^{2+} further increased (Wissenbach et al. 2000).

Swelling and stretch-activated channels that are nonselective for small monovalent cations and apparently impermeable to Ca^{2+} were found in the basolateral membrane of frog proximal tubule (Hunter 1990), in the mouse kidney cell line mIMCD-3 (Ono et al. 1994), and in human epididymal cells (Chan et al. 1994). Unitary conductances were in the range of 7 pS to 25 pS. In toad urinary bladder and in the A6 cell line derived from *Xenopus* kidney, cell swelling activated nonselective cation channels that were blocked by extracellular Ca^{2+} at millimolar concentrations (van Driessche and Erlij 1994; van Driessche et al. 1994). The actual role of these channels in the RVD process is completely unclear and remains to be elucidated.

Organic osmolytes in RVD

After exposure to hypotonic media, cells begin almost immediately to release organic osmolytes in addition to inorganic osmolytes. This release is not the consequence of a general increase of the membrane permeability to organic solutes, but shows a marked specificity (Grunewald and Kinne 1989; Furlong et al. 1991) and thus involves specific release pathways. These pathways are only briefly activated as exemplified for sorbitol efflux from IMCD cells in Fig. 5 (Kinne 1998).

Some of them have quite low osmotic thresholds. In some cells, organic osmolyte release accounts for about 50% of the total osmolytes released during RVD; thereby, its importance in overall cell volume regulation becomes evident. It also becomes necessary for the cells to coordinate the various channel activities so that they can operate in concert at the proper activity. How this coordination is achieved is one of the enigmas that remain to be solved.

Swelling-activated taurine release

Taurine release induced by cell swelling is a well-studied phenomenon in Ehrlich Ascites tumor cells (Hoffmann et al. 1988), articular chondrocytes (Hall and Bush 2001), HeLa cells (Hall et al. 1996; Stutzin et al. 1999), glial cells (Jackson and Strange 1993; Roy 1995; Roy and Malo 1992), astrocytes (Pasantes-Morales et al. 1994b; Mongin et al. 1999; Olson 1999), liver cells (Warskulat et al. 1997a), endothelial cells (Weik et al. 1998; Nilius and Droogmans 2001), neurons (Pasantes-Morales et al. 1994a), and renal medullary cells (Banderali and Roy 1992b; Ruhfus and Kinne 1996; Sanchez-Olea et al. 1991). During RVD, 30% to 50% of the intracellular taurine leave the cell. Taurine efflux is passive and directed only by the concentration difference (Sanchez et al. 1991; Roy and Malo 1992), lacks saturability and transstimulation. In addition, taurine flux in a variety of cells is inhibited by a range of anion channel blockers (Goldstein and Davis 1994; Boese et al. 1996b; Goldstein et al. 1996; Hall et al. 1996; Kinne et al. 1996; Ruhfus and Kinne 1996; Junankar and Kirk 2000). These properties suggest that taurine efflux is mediated by a transport system which is more similar to a channel than to a transporter (Kirk 1997; Kirk and Strange 1998; Perlman and Goldstein 1999; Shen et al. 2002). The search for such channels has revealed the following candidates thus far (see also Pasantes-Morales et al. 1994b).

VSOR-Cl^- channel, VSOAC, or VRAC

As discussed in more detail above (see the section entitled "Cl^- channels") these channels are outwardly rectifying (Boese et al. 1996a; Kirk and Strange 1998; Nilius and Droogmans 2001) and have significant permeabilities to anionic amino acids, including those with anionic side chains (glutamate and aspartate), as well as the anionic form of zwitterionic amino acids such as taurine. Attempts to further investigate the substrate specificity with regard to the organic osmolytes by measuring the effect of extracellular amino acids on RVD in glia cells (Pasantes-Morales et al. 1994b) or by electrophysiological studies revealed that the channel has a higher permeability to β-amino acids compared to α-amino acids. The anionic amino acids glutamate and aspartate appear to have a lower permeability than electroneutral amino acids of similar size (Pasantes-Morales et al. 1994b; Kirk 1997).

If compared to the chloride permeability of the channel, $P_{taurine}/P_{Cl^-}$ ranges from 0.15 to 0.25 in rat and human glial cells (Jackson and Strange 1993) and rat inner medullary collecting duct cells (Boese et al. 1996a), and to 0.49 in MDCK cells (Banderali and Roy 1992b). The size of the pore has been estimated to be between 5.6 Å and 11 Å (Jackson and Strange 1993; Nilius and Droogmans 2001). The channel therefore, seems to select the substrates both according to size and charge of the molecule.

Some electrophysiological studies also suggest that myo-inositol and sorbitol interact with the channel—since they are electroneutral molecules, no actual proof of translocation across the channel could be provided (Jackson and Strange 1993).

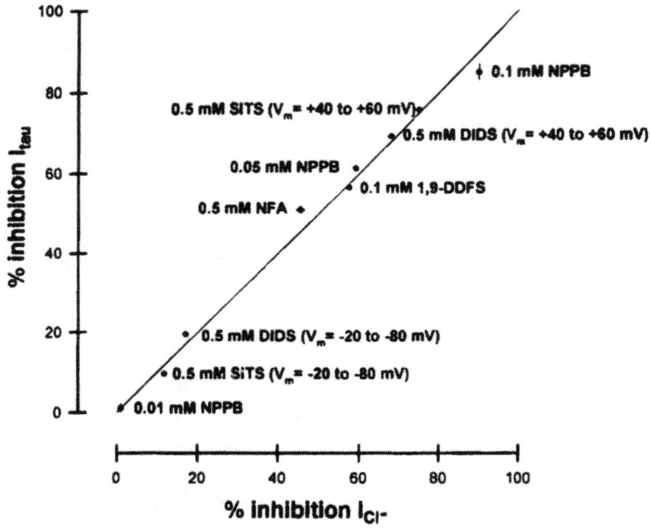

Fig. 6 Effects of various anion channel blockers on swelling-activated anion conductance in rat IMCD cells. Inhibition of volume-activated anion conductance under symmetrical taurine conditions (300 nM; pH 7.8; 31 nM charged; %inhibition I_{Tau}) plotted against inhibition of anion currents in symmetrical 140 mM CsCl conditions (%inhibition I_{Cl-}) by various anion channel blockers (unpaired experiments); n=5 for all experimental conditions. For most data points, SE values are smaller than the symbols used. *NFA* niflumic acid, *DDFS* 1,9-dideoxyforskolin, *NPPB* 5-nitro-2-(3-phenylpropylamino)benzoate, *DIDS* 4,4′-di-isothiocyanostilbene-2,2disulfonic acid, *SITS* 4-acetamino-4′-isothiocyanostilbene-2,2disulfonic acid (from Boese et al. 1996b)

In a variety of cells, VSOR-Cl^- channel and the taurine efflux pathways have an almost identical sensitivity to anion channel blockers. Figure 6 shows as an example the findings in rat IMCD cells. Because the rate of activation is also similar, these data suggest that chloride and taurine share the same release path.

Interestingly, VSOAC activity is regulated by the metabolic status of the cell. When intracellular ATP falls, the relative taurine efflux decreases. This has been observed in rat glial cells (Jackson et al. 1994), skate hepatocytes (Jackson et al. 1996), endothelial cells (Oike et al. 1994), and rat IMCD cells (Ruhfus and Kinne 1996). In patch clamp experiments, it was demonstrated that ATP or nonhydrolysable ATP derivatives have to be present at the cytoplasmic side of the membrane for activation of the channel by cell swelling (Jackson et al. 1994; Oiki et al. 1994; Oike et al. 1994; Jackson et al. 1996). These results point against a direct role of phosphorylation in the activation process. However, both phorbol esters and cAMP potentiate swelling induced organic osmolyte flux in C6 glioma cells, involving phosphorylation of VSOAC as a modifier of channel properties (Jackson and Strange 1993).

In agreement with findings of Motais (Motais et al. 1991) in trout red blood cells and in skate erythrocytes (Wittels et al. 2000) on swelling-activated taurine efflux, the activation of VSOAC is suppressed by high intracellular chloride concentrations (Jackson et al. 1996). The regulation by ATP and Cl^- has been interpreted to indicate that the cell very sparingly uses the metabolically expensive organic osmolytes only in situations where sufficient intracellular energy remains for a rapid recovery (Jackson et al. 1994; Jackson et al. 1996).

ICl$_n$ protein

The ICl$_n$ protein was cloned first from MDCK cells and exhibits a nucleotide-sensitive outwardly rectifying chloride current when expressed in oocytes (Krapivinsky et al. 1994; Fürst et al. 2000a). These properties are similar to those of swelling-activated chloride channels (see the section entitled "Cl$^-$ channels"). When reconstituted in lipid bilayers (Garavaglia et al. 2002) and in oocytes (Stegen et al. 2000), ICl$_n$ shows a significant permeability for taurine. Its proposed molecular structure composed of ß-barrels also bears similarities to porins invoked in organic solute transport in bacteria (Schirmer et al. 1995; Saier 2000). Furthermore, in neonatal myocytes, taurine efflux accompanies ICl$_n$ transposition from the cytoplasm to the plasma membrane with a similar time course, suggesting, but not proving, that ICl$_n$ is a regulator of the taurine channel or the channel itself. Knock down studies using antibodies or antisense oligonucleotides showed a significant reduction of RVD-related chloride currents (Gschwentner et al. 1995; Chen et al. 1999a); studies involving overexpression of ICl$_n$ showed the reverse (Hubert et al. 2000). Unfortunately, no data on organic osmolyte fluxes are available from these experiments. Thus, the role of ICln as one of the swelling-activated taurine channels remains to be determined. The potential role of the Cl$^-$/HCO$_3^-$ antiporter and of phospholemman as taurine channel is discussed in the section entitled "Cl$^-$/HCO$_3^-$ antiport" and the section entitled "Phospholemman," respectively.

Finally, it should be noted that in several cells separate pathways for chloride and taurine have been described; thus, additional transporters seem to exist (Shennan et al. 1994; Lambert and Hoffmann 1994; Stutzin et al. 1999).

Myo-inositol release pathways

Flux studies in rat C6 glioma cells (Jackson and Strange 1993) and rat inner medullary collecting duct cells (Ruhfus and Kinne 1996; Ruhfus et al. 1998) revealed striking similarities between swelling-activated myo-inositol and taurine fluxes. In glioma cells and IMCD cells, the rate of activation, the extent of release, and the sensitivity to anion channel inhibitors were virtually identical (Jackson and Strange 1993; Ruhfus and Kinne 1996). In addition, dependence on intracellular ATP levels could be demonstrated in the two cell types for both fluxes. In IMCD cells the majority of the taurine and myo-inositol occurs across the same membrane at the basal-lateral cell side (Kinne et al. 1996; Ruhfus et al. 1998). Furthermore, the activation pathways (Ca^{++}-independent and G-protein independent) are identical (Ruhfus and Kinne 1996). Electrophysiological investigations on the VSOAC in glia cells also showed that myo-inositol probably interacts with the channel (Jackson and Strange 1993).

Thus, it is very likely that myo-inositol efflux occurs via the same channel as taurine. Also, studies in skate erythrocyte suggest that taurine and myoinositol share a swelling-activated transport pathway (Goldstein and Davis 1994; Goldstein et al. 1996; Perlman and Goldstein 1999).

Sorbitol release pathways

Swelling-activated efflux of sorbitol has been observed in C6 glioma cells (Jackson and Strange 1993), a rabbit renal papillary cell line (Siebens and Spring 1989; Garty et al.

1991), HeLa Cells (Hall et al. 1996), rat IMCD cells (Grunewald and Kinne 1989; Ruhfus et al. 1998), rabbit kidney TALH cells (Kinne et al. 2000), and porcine urinary bladder cells (Mahler et al. 1998). In all instances tested, the flux, either measured as unidirectional sorbitol uptake (Jackson and Strange 1993; Hall et al. 1996; Kinne et al. 2000) or as solute efflux (Garty et al. 1991), lacked saturation kinetics. Furthermore, anion channel blockers such as NPPB, DIDS, 1,9-dideoxyforskolin, and cation channel blockers such as quinidine, inhibited sorbitol efflux in most systems, suggesting the involvement of a channel-like protein. With regard to the specificity towards the polyol, the length of the polyol chain and the orientation of the hydroxyl group at C6 appear to affect the permeability, although to a relatively small extent (Napathorn and Spring 1994).

In most cells, except IMCD cells, anions (chloride, iodide), potassium, choline, thymine, and taurine seem to be translocated via the same swelling-activated channel as sorbitol. This conclusion is based on the very similar inhibitory profile of various inhibitors on the fluxes (Hall et al. 1996), a similar lag phase before opening, and a similar dependence on the intracellular ATP level. In addition, in electrophysiological experiments in C6 cells sorbitol affected the volume-activated anion channel in such a way as to suggest an interaction with the anion binding site (Jackson and Strange 1993).

In rat IMCD cells, however, several differences between taurine/myo-inositol efflux and sorbitol efflux have been found. DIDS and SITS stimulate rather than inhibit sorbitol efflux; time of activation and closure of the two efflux pathways differ, as does the osmotic threshold for activation (Kinne 1998). Furthermore, activation of the sorbitol efflux pathway is calcium-dependent, probably involving activation of G-proteins (Ruhfus et al. 1996). Thus, for example, sorbitol efflux is reduced at low extracellular calcium and stimulated by arachidonic acid (Tinel et al. 1997; Kinne et al. 2000) and mastoparan under isotonic conditions (Ruhfus et al. 1996), whereas taurine (and myo-inositol) efflux is not affected by a change in overall calcium or by mastoparan, and inhibited by arachidonic acid (Kinne et al. 1996). However, local changes in the calcium concentration appear to affect swelling-induced Cl^- channels (Lemonnier et al. 2002).

It has to be noted, however, that in IMCD cells a considerable (about 50%) fraction of swelling-activated sorbitol efflux is not calcium dependent; thus, the existence of an additional channel—similar to the one proposed for other cells—cannot be excluded.

On the other hand, there is recent evidence for a highly selective sorbitol channel. Studies on malaria-induced pathways in chicken erythrocytes have shown that a solute channel is induced after infection with Plasmodium gallicaneum, which exhibits an about 17-fold higher permeability to sorbitol than to taurine and is relatively insensitive to DIDS and tamoxifen (Staines et al. 2002).

Other systems involved in volume regulation

Cl^-/HCO_3^- antiport

The superfamily of bicarbonate transporters (BTs) consists of two groups, namely, the AE family of anion exchangers (i.e., Cl^-/HCO_3^- antiporters) and the family of HCO_3^- transporters that are coupled to Na^+ (NBCs; see Boron 2001 for review). Whereas no direct contribution of NBCs to cell volume regulation has been reported so far, Cl^-/HCO_3^- antiport is of considerable importance for, both RVI as well as RVD. Four isoforms have been cloned to date (AE1–AE4) starting with murine AE1 (erythrocyte band 3) as early as 1985 (Kopito and Lodish 1985). The erythrocyte AE1 is a 929 amino acid protein with two functional domains, a hydrophilic one that is related to its role as an anchor for the cytoskeleton, and a hydrophopic one reflecting its function as an anion exchanger with 12 transmembrane regions (Kopito and Lodish 1985). An N-terminal truncated isoform of AE1 is present in the kidney (Brosius III et al. 1989). AE2 is widely expressed in nonerythroid tissues, including the kidney, and AE3 appears to be mainly expressed in electrically excitable tissues. Recently, AE4 was cloned from kidney; it appears to represent the apical Cl^-/HCO_3^- antiporter in β-intercalated cells (Tsuganezawa et al. 2001).

One of the prime functions of Cl^-/HCO_3^- antiport in cell volume regulation is the HCO_3^--dependent uptake of Cl^- that is induced parallel to the hypertonic activation of Na^+/H^+ antiport. Both transporters are functionally coupled via cell pH and the net result of their activity is a quasi electroneutral uptake of Na^+ and Cl^-. Activation of Cl^-/HCO_3^- antiport parallel to that of Na^+/H^+ antiport has been reported from a variety of system, including mammalian hepatocytes (Graf and Häussinger 1996), human retinal pigment epithelial and nonpigmented ciliary epithelial cells (Civan et al. 1994b; Civan et al. 1996), C6 glioma cells (Mountian et al. 1996), flounder red blood cells (Weaver et al. 1999), as well as from mouse medullary thick ascending limb (Hebert and Sun 1988). In addition, a parallel hypertonic stimulation of endogenous NHE1 and of heterologously expressed AE2 was reported from *Xenopus* oocytes (Jiang et al. 1997). A concerted activation of Na^+/H^+ antiport and Cl^-/HCO_3^- antiport was also found in renal epithelial (MDCK-F) cells that employ these mechanisms of RVI for the purpose of locomotion (Klein et al. 2000).

Equally well, Cl^-/HCO_3^- antiport may be activated under hypotonic conditions. In skate and trout erythrocytes, and also when heterologously expressed in *Xenopus* oocytes, AE1 appears to mediate the hypotonicity-induced release of Cl^- (see the section entitled "Swelling activated taurine release;" Musch et al. 1994; Fievet et al. 1995; Motais et al. 1997; Guizouarn et al. 2001). Of note, it may do so in a channel-like fashion and, at the same time, it also appears to increase cell membrane Na^+ and K^+ permeabilities (Guizouam et al. 2001).

AE 1 is one of the most clearly established molecular identities involved in taurine transport from fish red blood cells that are nucleated (Motais et al. 1991; Motais et al. 1997; Guizouarn et al. 2001). In particular, for AE 1 derived from trout, it has been convincingly demonstrated that expression of the protein in oocytes elicits both anion conductance and taurine transport, which show the same inhibitor profile and are clearly distinguishable from the endogenous $I_{Cl\ swell}$ of the oocytes (Fievet et al. 1995; Fievet et al. 1998; Koomoa et al. 2002).

The fact that in oocytes AE 1—contrary to intact trout red blood cells—cannot be activated by hypotonic stress and is not DIDS-sensitive might reflect different conformations of the protein in the two systems. Different conformations that change transport characteristics and sensitivity to modification have also been reported for ICl_n and phospholemman, and might be a general property of volume-regulated organic osmolyte channels. Another caveat remains: mouse AE 1 does not act as an osmolyte channel. In addition, in eel erythrocytes, significant differences between the AE 1-mediated SO_4 transport and swelling-activated taurine transport were observed (Lewis et al. 1996).Thus, a role of AE 1 in taurine release from mammalian cells remains to be demonstrated.

Phospholemman

Phospholemman (PLM) is a small membrane protein of 15 kDa which was first purified, cloned, and sequenced from dog heart (Palmer et al. 1991). The mature protein consists of 72 amino acids and has one single transmembrane domain. The name PLM denotes its high degree of phosphorylation, which (under most physiological conditions) appears to be mediated by protein kinases A and C (Palmer et al. 1991; Walaas et al. 1994; Mounsey et al. 1999). When expressed in *Xenopus* oocytes, PLM induces anion currents that are activated by large hyperpolarizations of membrane voltage and most likely it does so by inducing endogenous oocyte channels (Moorman et al. 1992; Mounsey et al. 1999). On the other hand, when reconstituted in lipid bilayers, PLM could be shown to form functional ion channels itself which exhibited rather slow gating kinetics and an enormous unitary conductance that is close to 700 pS (Moorman et al. 1995). It was also found that PLM (again in lipid bilayers) exhibits two different modes of operation, namely a cation- and an anion-selective one, and that the protein appears to spontaneously switch between these modes (Kowdley et al. 1997; Moorman and Jones 1998). PLM also exhibits a remarkably high permeability for taurine with a $P_{taurine}/P_{Cl}$ as high as 70 (Moorman et al. 1995) and a high selectivity with regard to the NH_2 group at the β-position (Moorman et al. 1995; Kowdley et al. 1997; Moorman and Jones 1998).

Overexpression of PLM in HEK293 cells significantly increased the membrane currents elicited by *hypotonic* stress (Moorman and Jones 1998) as well as the release of Cl^- (I^-) and taurine during RVD (Morales-Mulia et al. 2000); moreover, this osmolyte release could be markedly reduced by use of PLM-specific antisense oligonucleotides (Morán et al. 2001).

In solitary rat hepatocytes in primary culture, which could be shown to exhibit a distinct RVI (Kirschner et al. 1998) and which do express PLM, *hypertonic* conditions activate a channel that very much resembles PLM with respect to its unitary conductance, its gating pattern, and its nonselectivity for Na^+ over K^+. Moreover, in *Xenopus* oocytes expressing hepatocyte PLM, hypertonic stress induces a nonselective cation conductance and noise-analysis reveals the activation of a channel with characteristics that appear to be similar to those of PLM (Kirschner et al. 2003).

Taken together these results suggest a role of PLM in both RVD *and* RVI. This could mean that PLM functions as a last line of defense so that a cell under strong anisotonic conditions opens a large osmotic shunt just to survive. As an alternative view, given the regulatory role PLM appears to play in some systems (Moorman et al. 1992; Mounsey et al. 1999) and its high degree of phosphorylation per se (Palmer et al. 1991; Walaas et al. 1994; Mounsey et al. 1999) the protein may well be part of the signaling machinery that actually tunes the ion transporters responsible for cell volume regulation.

Na^+, K^+-ATPase

Cell volume homeostasis depends on the balance between osmolyte entry into the cell and removal of osmolytes from the cell interior. For cells that possess sodium entry pathways, such as sodium channels, Na^+/H^+ antiporter, Na^+, K^+, $2Cl^-$ symporter, or other sodium co-transporters, sodium entering the cells is removed by the primary active Na^+, K^+-ATPase. The activity of this enzyme, thereby plays an important role in maintaining the cell volume as indicated by the cell swelling observed in isotonic media when ouabain is present in the incubation medium. This observation reflects the actual stoichiometry of Na^+, K^+-ATPase of 3 Na^+/2 K^+ which continuously leads to a net loss of cellular inorganic osmolytes. In this context, it is also of note that in some systems an *inhibition* of the enzyme appears to directly contribute to the RVI processes (see below).

Since the first description of the enzymatic activity in 1957 by Skou (Skou 1957), the transport cycle has been described in great detail, the two subunits α and β have been identified, and a cell-specific distribution of their isoforms has been described (for review see Blanco and Mercer 1998).

The α-subunit has a molecular weight of 110 kDa, includes ten predicted transmembrane domains, and is responsible for the catalytic activity of the enzyme. During the pump's catalytic cycle the α-unit is transiently phosphorylated on a residue of the large cytoplasmic loop between transmembrane domains 4 and 5. This protein also contains phosphorylation sites for PKA and PKC. Residues of the fourth transmembrane segments contribute to the cation selectivity of the enzyme (for review see Dunbar and Caplan 2000).

The β-subunit has a molecular mass of approximately 65 kDa and elicits only one transmembrane segment with a cytosolic N-terminus. The extracellular C-terminus is heavily glycosylated (for review, see Dunbar and Caplan 2000). Assembly between the α- and β-subunits is required for the enzyme complex to exit the ER.

In addition, a γ-subunit has been identified; there is still considerable debate as to whether this subunit is required for the enzyme activity and/or whether it represents a regulatory membrane protein, which in some cells is closely associated with the enzyme (for review see Therien and Blostein 2000). The variability of the expression of the different subunits, the composition of which changes the properties of the enzyme considerably, results in a large variation of the response of the enzyme to changes in extracellular osmolality. In the context of this chapter, therefore, only a few examples of typical responses will be discussed.

During RVI following the exposure of cells to shrinkage in hypertonic media, the Na^+, K^+-ATPase is particularly important in cells that primarily and predominantly activate regulatory osmolyte transport systems, which enhance the intracellular sodium content, such as the Na^+/H^+ antiporter, the Na^+, K^+, $2Cl^-$ symporter, or a sodium channel, as discussed above. In a study in which osmolyte and Na^+ transport balances of rat hepatocytes were quantified as a function of hypertonic stress (see Fig. 7), it could be demonstrated that ouabain-sensitive ^{86}Rb uptake, representing the transport activity of the Na^+, K^+-ATPase, increased almost fourfold in a saturable fashion with increasing extracellular osmolarity (Wehner and Tinel 2000).

This increase could be completely accounted for by the rise in intracellular sodium concentration and the apparent Michaelis-Menten constant for sodium of the enzyme of 12 mMol/l.

The regulation of transport activity by changes of intracellular sodium represents the most straightforward and well-known response of the Na^+, K^+-ATPase to an osmotic challenge (Wojnowski and Oberleithner 1991). There are, however, also instances where either no increase in transport activity or even a decrease during RVI is observed. No change in ouabain-sensitive $^{86}Rb^+$ uptake was, for example, observed in the human hepatoma cell line Hep G2 (Wehner et al. 2002a). In view of the already relatively high Na^+, K^+-ATPase transport activity in these cells under isotonic conditions, the authors argued that the enzyme might normally function at V_{max} level and therefore a further stimulation might not

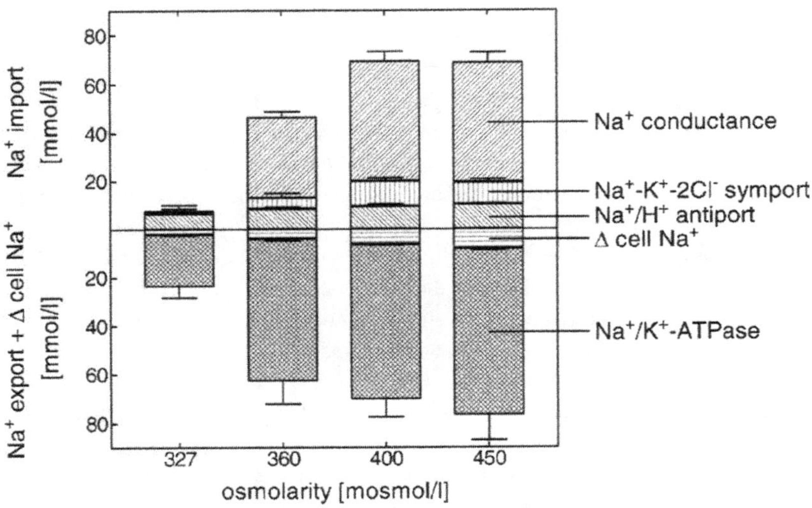

Fig. 7 Balance between Na^+ import and Na^+ export by the Na^+, K^+-ATPase plus the actual increase of cell Na^+. Rat hepatocytes in primary culture were exposed to increasing osmolarities. Note the saturability of the Na^+, K^+-ATPase pump rate (from Wehner and Tinel 2000)

be possible. In Malpighian tubule, an inhibition of Na^+, K^+-ATPase was observed during RVI. Here, the increase in intracellular adenosine, which inhibits the enzyme activity, was supposed to cause this effect (Caruso-Neves and Lopes 2000). In view of the fact that the Na^+, K^+-ATPase activity may also be altered by various kinases and phosphatases, as well as eicosanoids, intracellular ATP levels, intracellular pH (Goldstein and Davis 1994; Therien and Blostein 2000), rate of K^+ recycling across the membrane, and the membrane potential, the response of the enzyme during volume regulation can differ considerably depending on the cellular system being investigated.

It has to be noted that the alterations in Na^+, K^+-ATPase transport activity described above included regulation via phosphorylation, changes in transport rate, substrate, or inhibitor concentrations, or modifications in driving forces without changes in mRNA or enzyme content. Changes in the latter two parameters are observed when cells are exposed in vitro to hypertonic media for an extended time period. Thus, in mouse inner medullary collecting duct cells (Ohtaka et al. 1996), MDCK cells (Bowen 1992), human proximal tubule cells (Yordy and Bowen 1993), and vascular smooth muscle cells (Muto et al. 1998), changes in α-mRNA and β-mRNA are observed as early as one h after exposure of cells to a hypertonic medium. A maximum response is found between 8 h and 24 h of anisotonic incubation. The increase in mRNA leads to an augmentation of Na^+, K^+-ATPase abundance as demonstrated by a higher enzymatic activity in cell homogenates. An in vivo modulation of Na^+, K^+-ATPase expression was also observed in the renal papilla of water-deprived mice versus water-loaded mice, suggesting an osmoregulation of the enzyme in the intact animal (Capasso et al. 2001).

The mechanisms involved in the upregulation are still unclear; changes in the intracellular sodium concentration combined with an increased synthesis of intermediate regulatory proteins have been discussed (Yordy and Bowen 1993; Muto et al. 1998).

The physiological rationale for this long term adaptation is probably the same as for the short-term activation during RVI. In both instances, the cell aims to maintain a suitable sodium gradient across the plasma membrane. This mechanism assures the proper operation of sodium cotransport systems that include sodium osmolyte cotransport systems (see the section entitled "Organic osmolytes in RVI") and also systems regulating important intracellular parameters, such as intracellular pH and intracellular calcium. Thus, the activation of a net osmolyte export system during RVI, a reaction which at first sight appears to be counterproductive, fulfills an important function for general cell activity.

The effect of cell swelling on the Na^+, K^+-ATPase is difficult to estimate since a variety of stimulating and inhibiting parameters change simultaneously. Thus, cell swelling reduces intracellular sodium concentration, but activates K^+ channels (see the section entitled "K^+ channels") that facilitate K^+ recycling and thus could increase Na^+, K^+-ATPase activity. The increase in membrane potential would tend to decrease the transport rate of the electrogenic pump. The net effect of these—and other—regulatory changes differs from cell to cell. Nevertheless, there is evidence from various tissues that ouabain inhibits RVD, pointing to a significant role of the Na^+, K^+-ATPase in removing excess inorganic osmolytes from the cells (Fraser and Swanson 1994; Grunewald et al. 1994; Kanli and Norderhus 1998).

Signal transduction in volume regulation

Sensory systems and set points

Although the search for sensory mechanisms of cell volume changes has been carried out very intensively during recent years, a conclusive model of an osmosensor in mammalian cells has not yet been presented. The main cause for this gap lies in the nature of cellular signalling events. Usually, several signalling mechanisms are activated in parallel and some of them do influence each other mutually. This integrative character of signalling networks makes the assignment of signalling events to distinct pathways difficult. The concept that not one single osmosensor but rather a complex sensory system exists, has to be taken into account. Such a sensory system has to fulfill two major requirements. On the one hand, it has to be very sensitive to perceive even slight alterations of cell volume. On the other hand, the sensor must amplify and transfer the information to trigger signalling cascades leading to the activation of volume regulatory processes.

The mechanisms that have been reported to be implicated in sensing changes of cell hydration can be classified into three major groups, according to their cellular location. They can be (1) associated with the cytoskeleton perceiving changes in the interaction between extracellular matrix and plasma membrane receptors; (2) intracellular solute sensors sensitive to changes in macromolecular crowding and/or ionic strength and (3) membrane-based sensors which record alterations of cell membrane stretch.

Extracellular matrix and cytoskeleton

Cells adhere to the ambient extracellular matrix (ECM) via cellular integrins which are clustered to focal adhesions upon binding of ECM-ligands. Proteins that are associated with these adhesion sites may serve to maintain structural integrity, and equally well, they may be involved in signal transduction. For instance, the focal adhesion kinase FAK is a tyrosine kinase that is involved in the transduction of integrin-mediated cellular responses (Burridge and Zhong 1997). The activity of both integrins and FAK is altered by focal adhesion clustering.

Focal adhesions and the intracellular actin cytoskeleton effect each other mutually in their structures. Interestingly, evidence has been provided which indicates that integrins could be part of a volume-sensing system. In rat skeletal muscle, for example, the ECM-integrin-mediated cell adhesion and the cytoskeleton are involved in mechanochemical transduction of cell volume changes to chemical signals modulating glutamine transport (Low and Taylor 1998). ECM-integrin binding also transduces changes in cell volume and/or shape into changes in endothelial permeability of cultured bovine aortic endothelia and frog mesenteric capillaries (Kajimura et al. 1997). At the frog neuromuscular junction, integrins contribute to the hypertonic enhancement of transmitter release (Kashani et al. 2001).

The finding that integrins may be involved in sensing cell volume changes is coincidental with the circumstance that in nearly all systems studied so far volume changes cause alterations of the cytoskeletal organization. In the majority of the systems examined, hypotonic cell swelling evokes disorganization of the F-actin cytoskeleton. This has been shown for Ehrlich Ascites tumor cells (Pedersen et al. 2001), rat IMCD cells (Czekay et al. 1994) shark rectal gland and C6 glial cells (reviewed in (Henson 1999; Papakonstanti et al. 2000; Pedersen et al. 2001). In contrast, hypotonic actin polymerization could be detected in rat hepatocytes (Theodoropoulos et al. 1992) and HL-60 cells (Hallows et al. 1991). Increasing extracellular tonicity enhances the relative F-actin content in Ehrlich Ascites tumor cells (Pedersen et al. 2001) and lowers it in HL-60 cells (Hallows et al. 1991). These examples illustrate how diverse the response of the actin cytoskeleton to cell volume changes may be.

The alteration of cell volume evokes changes of cell morphology which are paralleled by the development of mechanical tension. This tension impinges on both the cytoskeleton and the interaction between integrins and their ECM-ligands. In both cases, integrins are altered in their activity by conformational changes. They may be activated and clustered to focal adhesions or inactivated and dispersed from focal adhesions (Burridge and Chrzanowska-Wodnicka 1996; Ingber 1997; Schoenwaelder and Burridge 1999).

The assembly of integrins to focal adhesions can influence the cytoskeleton by two synergistic pathways, both of them involving activation of the small GTPase Rho. The contribution of Rho to the regulation of volume-sensitive transport mechanisms has been reported for several systems, (see Nilius et al. 1999; Okada 1997).

Activation of RhoA by integrin-mediated adhesion (Renshaw et al. 1996; Clark et al. 1998; Ren et al. 1999) and/or by isometric tension of cytoskeletal components (Burridge and Chrzanowska-Wodnicka 1996) induces the formation of stress fibers (Amano et al. 1997). This change is caused by influencing the state of contractile structures or by effecting the conformation of actin-associated proteins (Schoenwaelder and Burridge 1999). Rho activates Rho-kinase, which phosphorylates and thereby inhibits myosin phosphatase. This results in elevated phosphorylation of myosin light chain (MLC; Kimura et al. 1996). The consequence is that myosin filament assembly and actin-activated myosin ATPase activity (Burridge and Chrzanowska-Wodnicka 1996) are enhanced. This leads to the bundling of actin filaments and the assembly of stress fibers. Rho may also phosphorylate MLC directly (Amano et al. 1996).

The alteration of stress fiber assembly by actin-associated proteins occurs via lipid derived messengers. The activation of PI-5 kinase by RhoA elevates the intracellular level of phosphatidylinositol 4,5-bisphosphate (PIP_2; Chong et al. 1994). PIP_2 induces conformational changes of the actin-associated proteins vinculin (Weekes et al. 1996), ezrin-radixin-moesin (ERM; Niggli et al. 1995), and profilin. Thereby, binding sites for actin and other proteins are exposed. The resulting changes in stress fiber tension are then transduced to the integrins, which thereby are altered in their activity. This bidirectional signalling between the cytoskeleton and integrins allows the creation of a feedback-loop, by which the duration of volume-generated signalling can be determined (Burridge and Chrzanowska-Wodnicka 1996; Schoenwaelder and Burridge 1999).

In this context, another property of the sensor of volume changes, namely, the transduction of the perceived information to effectors, may be realized by three different mechanisms. Firstly, alterations in the cytoskeletal structure may affect the activity of signalling molecules by changing their intracellular localization. Secondly, the cytoskeleton may directly interact with and act upon volume regulatory transporters and. thirdly, vesicular transport, if involved in cell volume regulation may be influenced, (see for example Czekay et al. 1994).

In many studies cited above, it is not yet clear whether the perturbations of the cytoskeleton observed are the cause or the consequence of volume changes and/or regulatory processes. Furthermore, a direct interaction with the target proteins and the consequences of this interaction have not been recorded in the intact cell. Thus, a higher spatial and temporal resolution of these processes is necessary to be able to conclusively judge upon the role of the cytoskeleton and the ECM in volume sensing.

As was outlined above, the alteration of cell adhesion effects FAK, as well as integrin signalling. For the activation of volume-sensitive Cl^- currents, it has been proposed that the stimulation of a tyrosine kinase is one of the early events in osmosensing (Sadoshima and Izumo 1993; Tilly et al. 1993). In most instances, however, tyrosine-phosphorylarions last only for a short period of time, which is mainly due to the presence of tyrosine-specific protein phosphatases. Accordingly, these tyrosine-phosphorylations, in most systems, are translated into longer lasting signalling events, namely, by the activation of serine/threonine phosphorylation cascades, such as the mitogen-activated protein kinase (MAP kinase) pathways. The sequential patterns by which MAP kinases are activated offer the opportunity of an amplification, stabilization, and integration of intracellular signalling events in response to extracellular stimuli at the same time. In fact, MAP kinases could be shown to be involved in the regulation of a variety of fundamental physiological processes such as cytoskeletal organization, nuclear transcription, protein translation, cell growth and differentiation, as well as cell volume homoeostasis.

Activation of ERK1/2 MAP kinases by cell volume changes appears to be a quite common phenomenon. A swelling-induced activation occurs in most mammalian cell types (van Der Wijk et al. 2000). In contrast, activation of ERK1/2 upon hypertonic cell shrinkage was observed, for instance, in MDCK cells (Itoh et al. 1994), H4IIE rat hepatoma cells (Schliess et al. 1996), and in bovine aortic endothelial cells (Duzgun et al. 2000). In these systems, hypertonic activation occurs within 10 min and the maximal induction is reached after 20 min–30 min.

The activity of p38 MAP kinase can be upregulated by hypertonicity as well as by hypotonicity (see Shrode et al. 1998). Hypertonic activation occurs, for example in U 937 and in IMCD-3 cells. In the latter system, the activation of p38 is achieved after approximately 10 min (Zhang and Cohen 1996; Shrode et al. 1997). Hypotonic activation of p38 has been observed in the perfused rat liver (Vom Dahl et al. 2001), in HTC rat liver hepatoma cells (Feranchak et al. 2001), and in intestine 407 cells (van Der Wijk et al. 1998; van Der Wijk et al. 2000). For the perfused rat liver and intestine 407 cells it could be demonstrated that p38 is activated rapidly and without a detectable lag phase. In IMCD-3 cells, the JNK MAP kinase is activated under hypertonic conditions (van Der Wijk et al. 1998). The swelling-induced activation of JNK is a slow process, occurring with a delay of some 10 min–20 min (van Der Wijk et al. 2000).

Though MAP kinases are activated by the anisotonic alteration of cell volume in many systems, a direct involvement in volume regulatory mechanisms has only been observed in a few systems. In human cervical cancer cells, for example, swelling-activated ERK1/2 is linked to the activation of Cl^- currents, K^+ currents and taurine transport (van Der Wijk et al. 2000; Shen et al. 2001). In contrast, even though Na^+/H^+ antiport can be activated in an ERK1/2-dependent manner, the osmotic activation of this antiporter does not involve ERK1/2 (or p38 MAP kinase) in Chinese hamster lung fibroblasts (Aharonovitz and Granot 1996; Bianchini et al. 1997; van Der Wijk et al. 2000). Moreover, Nilius et al. demonstrated that hypotonic activation of MAP kinases are not involved in the activation of volume-sensitive Cl^- currents (Nilius et al. 1997).

In most instances, the apparent absence of effects of MAP kinases on the (fast) regulatory movements of inorganic osmolytes can be explained in terms of their slower time course of activation that instead reflects that of organic osmolyte transport and metabolism. Accordingly, osmosensitive MAP kinases may be related to cell volume regulation by activating transcription factors, since the required upregulation of the corresponding metabolizing enzymes and organic osmolyte transporters is realized by enhancing their levels of transcription. An adjusted gene expression thus allows the cells to sustain cell volume disturbances for a prolonged period of osmotic stress.

Macromolecular crowding and ionic strength

Macromolecules occur in a high intracellular concentration; they occupy typically 20%–30% of the total cell volume (Ellis 2001). This noncytosolic and not osmotically active space (see the section entitled "The osmometric behavior of cells") excludes molecules of low molecular weight. This phenomenon, known as macromolecular crowding, has fundamental effects on the rates and equilibria of biochemical reactions (Ellis 2001). Under conditions of increased macromolecular crowding, the reduction of excluded volume causes a decrease of total free energy of the solution. Therefore, the equilibrium constants of binding and association reactions that exclude the least volume to all the other macromolecules present are increased. Macromolecular crowding alters biochemical reaction rates by influencing both molecular diffusion rates and thermodynamic activities. While molecular diffusion rates will be reduced by increased crowding (Luby-Phelps 2000), their thermodynamic activity will be enhanced. The net result of these opposing effects depends on the nature of the rate-limiting step. If a reaction rate is limited by the encounter rate of the initial molecules, the rate will be decreased by crowding due to decreased diffusion rates. If the rate is limited by the activity of the transition complex, it will be increased by crowding due to enhanced thermodynamic activities (Ellis 2001).

Macromolecular crowding is reduced by increasing ionic strength, because proteins interact with the ambient electrolytes. The volume regulatory set points are thereby shifted to smaller volumes (Parker et al. 1995). Since even slight changes in macromolecular crowding are sufficient to evoke large changes in enzymatic activity, macromolecular crowding has been discussed as a candidate for a cell volume sensory system. Indeed, the volume sensitive transporters Na^+/H^+ symporter (Parker et al. 1995), K^+-Cl^- symport (Minton et al. 1992; Minton 1994; Parker et al. 1995), VSOAC channel (Emma et al. 1997), and Cl^- channels (Hoffmann and Pedersen 1998; Nilius et al. 1998) are influenced by macromolecular crowding and ionic strength in various systems. A reduction of ionic strength activates some transport systems involved in RVD. An increase in ionic strength is supposed to mediate the induction of sodium organic osmolyte transporters and osmosensitive enzymes during prolonged exposure of cells to hypertonic media (see Burg 1995; and the section entitled "Organic osmolytes in RVI").

Macromolecular crowding and ionic strength are quite global parameters of idealized cells that can not yet be related to various cellular compartments and the place where signal generation and transduction occurs. Thus, their importance in localized reactions remains to be determined.

Cell membrane stretch

Cell swelling causes stretching of the plasma membrane, whereas shrinkage leads to a decrease in cell membrane stretch (Kinnunen 2000). These alterations of membrane tension may generate a signal perceived by stretch-sensitive channels. Among the stretch-sensitive channels, both stretch-activated and stretch-inactivated channels can be found. Stretch-activated channels can either be nonselective cation channels permeable for K^+, Na^+ and Ca^{2+}, or channels selective for Ca^{2+}, K^+ or Cl^- (Hoffmann and Dunham 1995; Lang et al. 1998a).

The activation of *nonselective* stretch-activated cation channels upon cell swelling was demonstrated in several cells (Christensen 1987; Lansman et al. 1987; Falke and Misler 1989; Bear 1990; Christensen and Hoffmann 1992). Activation of these channels contributes to RVD, probably mainly by the resulting Ca^{2+} current (Hoffmann and Dunham 1995; Lang et al. 1998a). The resulting (local) increase in intracellular Ca^{2+} then activates Ca^{2+}-sensitive K^+ channels (Christensen 1987; Ubl et al. 1988). The loss of K^+ is accompanied by osmotic water movement, leading to volume decrease (see the sections entitled "K^+ channels" and "Cation channels").

The activation of *selective* stretch-activated ion channels contributes directly to cell volume regulation. Stretch- and volume-sensitive K^+ channels were demonstrated for *Necturus* proximal tubule cells (Filipovic and Sackin 1992), gall bladder cells (Vanoye and Reuss 1999) and enterocytes (Dubinsky et al. 2000). Similarly, cell membrane stretch may also lead to the activation of stretch-sensitive maxi Cl^- channels, which were reported to be activated upon cell swelling in several systems (see the section entitled "Cl^- channels").

Stretch-inhibited channels are not as common as stretch-activated channels, and their relevance for cell volume regulation is usually regarded as negligible (Morris 1990; Hoffmann and Dunham 1995). Nevertheless, Suzuki et al. showed, that a cation channel cloned from rat kidney is inactivated by membrane stretch due to cell swelling and activated by cell shrinkage (Suzuki et al. 1999).

The mechanism by which stretch-sensitive channels are regulated by cell volume changes has not yet been fully elucidated. Signals modulating stretch-sensitive channels could be mechanical tension, conveyed by the membrane or the cytoskeleton (Sokabe et al. 1991; Ingber 1997), or the release of stretch-sensitive messengers, such as fatty acids (Kirber et al. 1992), or ATP (Wang et al. 1996).

Though a variety of ion channels were reported to respond to membrane stretch caused by cell volume alterations, the physiological significance of this signal remains questionable, since the activation of these channels requires considerable stretch (Okada 1997). For example, it could be demonstrated that significant cell membrane stretch only arises when cell volume increases by 60%; volume regulatory mechanisms are, however, already activated when cell volume is altered by 5%–10% (Al Habori 2001). Thus, stretch-sensitive channels may only play a role in sensing cell volume changes when excessive cell swelling occurs and other signalling systems have failed (Lang et al. 1998a).

Lipid derived messengers

Phosphoinositides

Lipid signalling through phosphoinositides was reported to be involved in cell survival, cytoskeletal remodeling, metabolic control, and vesicular trafficking (Wymann and Pirola 1998). Since all these processes are related to cell volume regulation (Lang et al. 1998a; Hoffmann and Mills 1999), it seems not surprising that there is ample evidence for the involvement of lipid-derived messengers in processes maintaining osmolyte balance. Phosphoinositides can interfere at different levels with signalling cascades. As signalling intermediates, they can serve as substrate and cofactor for phospholipases or phosphatidylinositol kinases. They are also capable of regulating various enzymes like Rho GTPases and profilin and modulating ion channels and transporters.

For Ehrlich Ascites tumor cells, it could be shown that the level of phosphatidylinositol 4,5-bisphosphate [PI(4,5)P_2] responds to cell volume alterations. A decrease and an increase in cellular PI(4,5)P_2 levels occurs upon cell shrinkage and cell swelling, respectively (Hoffmann and Mills 1999). Similarly, a PI(4,5)P_2 decrease could be shown for rat hepatocytes (Baquet et al. 1991; Vom Dahl et al. 1991) upon cell shrinkage. In Swiss 3T3 mouse fibroblasts, hypertonic stress increases the levels of phosphatidylinositol 3,4-bisphosphate [PI(3,4)P_2] and phosphatidylinositol 3,4,5-trisphosphate [PI(3,4,5)P_3; Van der Kaay et al. 1999].

The effects of cell volume changes on PI(3,4,5)P_3 levels, phosphatidylinositol 3-kinase (PI3 K) and phosphatidylinositol 5-phosphatase activity, and the initiated signalling pathways are very variable. In rat skeletal muscle, hypotonicity as well as hypertonicity induce changes of glutamine uptake that are dependent on the activity of PI3 K (Low and Taylor 1998). The authors assume that PI3 K may serve to maintain an integrin-dependent mechanochemical transduction mechanism in an active state. The activity of PI3 K is also required to initiate the hypertonic activation of Na^+ conductance in rat hepatocytes (Wehner et al. 2002b).

In MDCK cells, hyperosmolality activates the serine-threonine-kinase Akt (also known as protein kinase B) in a PI3K-dependent manner (Terada et al. 2001). The product of PI3 K, PI(3,4,5)P_3, interacts with Akt itself as well as with 3-phosphoinositide-dependent kinases (PDK1 and PDK2). The actual activation of Akt occurs via Thr and Ser phosphorylation by PDK1 and PDK2 (Alessi et al. 1997; Stephens et al. 1998).

It was reported that the activation of Akt by hypertonicity prevents hyperosmolality-induced apoptotic changes in MDCK cells (Terada et al. 2001). Together with the finding that the hypotonic activation of PI3 K and Akt in HepG2 cells stimulates proliferation (Kim et al. 2001) ,this suggests that both enzymes influence the osmo-sensitive balance between cell survival and apoptosis.

In Swiss 3T3 mouse fibroblasts, hypertonicity-activated PI3 K elevates PI(3,4,5)P_3 levels without enhancing Akt activity, due to negative regulation of Akt by osmotic stress (Van der Kaay et al. 1999). This is coincident with the hypertonic, PI(3,4,5)P_3-dependent inhibition of PKBα in HEK-293 cells caused by inhibition of phosphorylation and activation of dephosphorylation of Akt, probably via PP2A (Meier et al. 1998). The inhibition of Akt seems to be triggered by PI(3,4,5)P_3 levels, independent of whether these occur by activation of PI3 K (Van der Kaay et al. 1999) or by inhibition of phosphatidylinositol 5-phosphatase (Meier et al. 1998).

Hypertonic accumulation of PI(3,4,5)P_3 activates p70 S6 kinase in Swiss 3T3 cells (Van der Kaay et al. 1999). This activation of p70 S6 kinase could be related to cell volume regulation on the basis of alterations in glycogen metabolism, since p70 S6 kinase activity is required for osmotically triggered changes in glycogen synthesis in rat skeletal muscle; during RVI, glycogen synthesis is inhibited, whereas it is stimulated during RVD (Low et al. 1996). The effect of cell volume changes on glycogen metabolism has also been demonstrated for rat hepatocytes (Häussinger et al. 1991). In this system, hypotonicity activates both PI3 K and p70 S6 kinase, but only PI3 K is involved in the hypotonic regulation of glycogen synthase (Krause et al. 1996).

The transcription of the serum and glucocorticoid dependent kinase (sgk) is clearly sensitive to cell volume changes. In the human hepatoma cell line HepG2, upregulation of sgk occurs upon cell shrinkage, whereas swelling reduces sgk transcript levels (Waldegger et al. 1997). Furthermore, sgk, like Akt and p70 S6, is a downstream target of PI3 K (Kobayashi and Cohen 1999). Most interestingly, stimulation of sgk results in the activation of ENaC in A6 cells (Chen et al. 1999b), rabbit cortical collecting duct cells (Naray-Fejes-Toth et al. 1999), and in *Xenopus* oocytes (Böhmer et al. 2000). For the latter system, it could also be demonstrated that sgk increases the abundance of the ENaC-protein in the plasma membrane (Alvarez et al. 1999; Wagner et al. 2001). In addition, sgk stimulates the activity of the neuronal K^+ channel Kv 1.3 as well as that of Na^+-K^+-ATPase and may thereby contribute to cell volume regulation (Setiawan et al. 2002; Wärntges et al. 2002).

The phosphorylation sites at which Akt and other protein kinases, such as p70 S6 kinase, PKCδ, and SGK are activated exhibit high similarity. This suggests that PDK1 and PDK2 may contribute equally well to the PI3K-dependent activation of these enzymes (Kobayashi and Cohen 1999).

Moreover, in rat HTC hepatoma cells (Feranchak et al. 1998) and in Mz-Cha-1 human cholangiocarcinoma cells (Feranchak et al. 1999), PI3 K contributes to cell volume regulation by the activation of Cl^- secretion via ATP release. The hypotonic activation of Na^+/H^+ exchange and HCO_3^- absorption in rat renal medullary thick ascending limb (MTAL) requires PI3 K activity as well (Good et al. 2000). The fact, that PI3 K is activated by *hypotonicity* as well as by *hypertonicity* in several systems reflects the general role of PI3 K as a key enzyme in signalling cascades. The actual result of this activation is then defined by its downstream targets that may vary considerably between different cells.

Arachidonic acid and eicosanoids

Arachidonic acid (AA) can be directly released from phosphatidylcholine and phosphatidylethanolamine by the activity of phospholipase A_2 (PLA_2). The cleavage of phosphatidylinositol by phospholipase C yields arachidonic acid out of diacylglycerol (DAG). Phospholipase D produces phosphatic acid, which can be transferred to arachidonic acid by the action of diglyceride lipase. Phosphatic acid may also be hydrolysed by PLA_2 into AA and lysophosphatic acid (LPA).

An increase in the intracellular concentration of arachidonic acid and its derivates upon hypotonic cell swelling has been shown for rat IMCD cells (Tinel et al. 1997), human neuroblastoma cells (Basavappa et al. 1998), and Ehrlich Ascites tumor cells (Thoroed et al. 1997). This release of AA seems to be mainly caused by the activation of PLA_2, as shown for Ehrlich Ascites tumor cells (Thoroed et al. 1997), human platelets (Margalit et al. 1993a), and human neuroblastoma cells (Basavappa et al. 1998).

In agreement with these findings, inhibition of PLA_2 prevents the activation of hypotonically activated events like Cl^- current in bovine pigmented ciliary epithelial cells (Mitchell et al. 1997) and iodide efflux in rat brain endothelial cells (von Weikersthal et al. 1997).

Influence on volume regulatory mechanism can be carried out by AA itself or by eicosanoids made from AA. AA is able to directly inhibit volume-sensitive Cl^- currents in guinea pig antral gastric myocytes (Xu et al. 1997), rat osteoblast-like cells (Gosling et al. 1996), Ehrlich Ascites tumor cells (Hoffmann and Lambert 1994), human neuroblastoma cells (Basavappa et al. 1998), and rat hepatocytes (Sakai et al. 1996). The activation of C1C-2 Cl^-channels by AA has been reported for the human lung epithelial cell lines Calu-3, A549, and BEAS-2B (Cuppoletti et al. 2001), and for HEK-293 cells (Tewari et al. 2000). AA is capable of activating K^+ channels in the colonic secretory cell line T84 (Devor and Frizzell 1998), in rat neuronal cells (Kim et al. 1995), and in rabbit cortical collecting tubule (CCT) cells (Ling et al. 1992).

Via alteration of cytosolic Ca^{2+} (Tinel et al. 1997; Jorgensen et al. 1999), AA can influence Ca^{2+}-sensitive Cl^- and K^+ currents, as well as sorbitol efflux in IMCD cells (see the section entitled "Sorbitol release pathways").

The variety of the response of organic osmolyte release pathways in IMCD cells to AA is illustrated in Fig. 8. Here, isotonic taurine and *myo*-inositol efflux is inhibited, whereas AA can mimic the hypotonic response for sorbitol, betaine, and GPC.

AA can be metabolized in several pathways; two of them have been reported to be involved in mediating mechanisms of cell volume regulation. The lipoxygenase pathway leads to formation of leukotrienes and prostanoids, whereas prostaglandins and thromboxanes are formed in the cyclooxygenase pathway by the activity of cyclooxygenase and peroxidase.

Products of the lipoxygenase pathway seem to be involved mainly in the activation of RVD. The 5-LIP metabolite LTD_4 is able to activate hypotonic Cl^- and K^+ currents in Ehrlich Ascites tumor cells (Jorgensen et al. 1996); the isotonic cell shrinkage induced by LTD_4 occurs by the activation of the same mechanisms (Lambert 1989). In addition, LTD_4 acts as a second messenger for the taurine leak pathway and thereby contributes to RVD in this system (Lambert and Hoffmann 1993). An alteration of K^+ channels by 5-LIP product has also been shown for mudpuppy red blood cells (Light et al. 1997). In human skin fibroblasts, the activity of 5-LIP is required for the activation of swelling-induced Cl^-

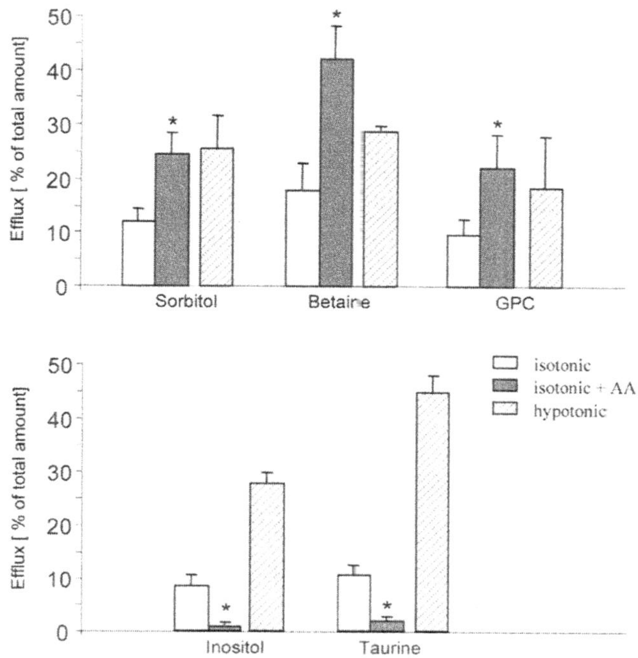

Fig. 8 Effect of arachidonic acid (*AA*) and hypotonicity on osmolyte efflux in rat inner medullary collecting duct (IMCD) cells. Efflux of osmolytes (in % of total amount; ±SE/10 min) was determined after 10 min exposure to 10 μM AA at isotonicity or after a decrease of extracellular osmolarity (600→300 mOsm/l) by means of HPLC; *n*=4 for all experimental conditions (B. Ruhfus and R.K. Kinne, unpublished data)

efflux (Mastrocola et al. 1993). The 12-LIP product hepoxilin A_3 is involved in the hypertonic activation of K^+ channels in human platelets (Margalit et al. 1993b).

In Ehrlich Ascites tumor cells, the stimulation of leukotriene synthesis is concomitant with an inhibition of prostaglandin synthesis (Lambert et al. 1987). The acceleration of RVD by LTD_4 or AA and the deceleration by prostaglandin E_2 (PGE_2) in mouse colonic crypts (Mignen et al. 1999) confirms that the reciprocal alteration of lipoxygenase and cyclooxygenase pathway affect synergistically volume recovery in these systems. In Ehrlich Ascites tumor cells, PGE_2 has no effect on the swelling-induced Cl^- permeability or K^+ permeability (Lambert et al. 1987). However, since PGE_2 is able to activate Na^+ channels, a decrease in PGE_2 synthesis may contribute to volume decrease by a reduction in the passive membrane conductance for Na^+ (Hoffmann 1978). In MDCK cells (Steidl et al. 1991) and human erythrocytes (Li et al. 1996), PGE_2 has been reported to activate K^+ channels resulting in a decrease of cell volume. Cl^- channels can be activated by cyclooxygenase-dependent PGE_2 synthesis in the hepatoma cell line HTC (Kilic and Fitz 2002). For human amnion-derived WISH cells, it could be shown that COX-2 gene expression is upregulated by hypotonic stress; as a consequence thereof, PGE_2 release is increased (Lundgren et al. 1997). Of note, there is no evidence for the involvement of PLA_2-signalling in cell volume restoration upon hypertonic stress.

An involvement of phospholipase C (PLC) in cell volume regulation has been reported after hypotonic cell swelling, as well as after hypertonic cell shrinkage. PLC can influence volume regulatory processes by different mechanisms, depending on the products of the reaction.

Ca^{2+} oscillations can be IP_3– as AA-sensitive as well (Ling et al. 1992; Tinel et al. 1997); both messengers result from PLC activity. Involvement of PLC in hypotonic Ca^{2+} oscillations has been reported for rat IMCD cells (Tinel et al. 1997), rat caudal artery endothelial cells (Shinozuka et al. 2001), rabbit proximale tubule cells (O'Neil and Leng 1997), and bovine aortic endothelial cells (Oike et al. 2000). The relevance of Ca^{2+} signalling for cell volume regulation is discussed in detail in the section entitled "Intracellular calcium."

In pigmented ciliary epithelial cells, volume-sensitive Cl^- currents are dependent on the activity of both PLA_2 and PLC. The contribution of PLC may be the synthesis of AA or enhancement of PLA_2 activity via a PKC-dependent phosphorylation of PLA_2 (Mitchell et al. 1997).

Phosphatic acid and lysophosphatic acid

Phospholipase D (PLD) generates the lipid second messengers phosphatic acid (PA) and choline by the cleavage of phosphatidylcholine. Only few examples exist for the involvement of PLD in volume-regulatory signalling. In skate erythrocytes, hypotonic treatment increases the turnover of phosphatidylcholine and the formation of PA due to the activation of PLD (Musch and Goldstein 1990). Sorbitol-induced osmotic stress enhances activity of PLD and PA release in rat adipocytes, leading to activation of aPKCs and translocation of GLUT4 transporters (Sajan et al. 2002). The PIP_2-synthesizing enzyme PIP 5-kinase can be regulated by PA; therefore, PLD may play a signalling role in PIP_2 sensitive events (Cockcroft 2001). Since alterations of PIP_2 levels are related to cell volume regulation, PLD may contribute to volume-dependent phosphoinositide signalling.

PA can be metabolized to AA by the action of PLD, but the relevance of PLD for volume-induced AA metabolism seems to be minor. In Ehrlich Ascites tumor cells, the contribution of PLD to the hypotonic AA release could be excluded (Thoroed et al. 1997).

Enzymatic conversion of PA by extracellular PLA_2 yields lysophosphatic acid (LPA); the initialization of intracellular signalling cascades occurs by binding of LPA to G protein coupled receptors (Goetzl and An 1998). For Ehrlich Ascites tumor cells, it could be revealed that extracellular application of LPA caused cell shrinkage by evoking Ca^{2+} oscillations and activation of Ca^{2+}-sensitive K^+ and Cl^- currents. As a consequence of this volume decrease, the Na^+/H^+ exchanger is activated (Pedersen et al. 2000). On the basis of the similarity between cell signalling after hypertonic cell shrinkage and after LPA treatment, the authors assume an interrelation in the regulation of both events.

Diacylglycerol

In erythrocytes, the intracellular level of diacylglycerol (DAG) is increased by hypotonicity (Musch and Goldstein 1990). DAG is a product of PLC and able to activate protein kinase C (PKC). PKC is a serine/threonine kinase of central importance in intracellular signalling cascades; hence, it is not astonishing that PKC is activated upon hypotonic cell swelling as well as upon hypertonic cell shrinkage in various systems.

Contribution of PLC to RVI after hypertonic stress has been reported for NIH/3T3 mouse fibroblasts (Zhuang et al. 2000) and Ehrlich Ascites tumor cells (Jensen et al. 1993); in both systems, PLC acts via PKC. In NIH/3T3 cells, the hypertonic phosphorylation of ERK 1/2 is due to DAG-dependent activation of classical and novel PKCs (Zhuang et al. 2000). Hypertonic activation of PLC enhances $Na^+/K^+/2Cl^-$ cotransport activity in Ehrlich Ascites tumor cells via Ca^{2+}-dependent PKC (Jensen et al. 1993).

In Ehrlich Ascites tumor cells, hypertonicity-activated PKC contributes to RVI by activating NHE and NKCC, and by inhibiting volume-sensitive Cl^- currents (Larsen et al. 1994; Hoffmann and Dunham 1995; O'Donnell et al. 1995; Pedersen et al. 1996). Inhibition of volume-sensitive Cl^- currents by PKC has also been reported for human nonpigmented ciliary epithelial cells (Coca-Prados et al. 1996), for cardiac ClC-3 expressed in NIH/3T3 cells (Duan et al. 1997), and HeLa cells (Hardy et al. 1995); however, in other systems, PKC may either stimulate chloride currents or leave them unaffected (reviewed in Pasantes-Morales et al. 2000a). The contribution of PKC to shrinkage-induced activation of NKCC is also cell-specific. While NKCC is downregulated by PKC in sweat glands (Toyomoto et al. 1997), NKCC remains unaffected in bovine endothelial cells (O'Donnell et al. 1995). PKC-dependent activation of NKCC has been demonstrated for rat hepatocytes; in this system, hypertonic Na^+ currents are activated by PKC, while hypertonicity activated NHE is not regulated by PKC (Heinzinger et al. 2001). In NIH/3T3 cells, hyperosmolality enhances the activity of c and n PKC, leading to activation of ERK1/2 (Zhuang et al. 2000).

After osmotic cell swelling, PKC is activated in the late phase of RVD in Ehrlich Ascites tumor cells; inhibition of VSCC by PKC terminates the volume response (Larsen et al. 1994). In contrast, hypotonic Cl^- membrane permeability is enhanced by PKC in HTC rat hepatoma and Mz-ChA-1 human cholangiocarcinoma cells (Roman et al. 1998).

The swelling-induced release of organic osmolytes is also influenced by the hypertonic activation of PKC. While inhibition of PKC prevents the hypertonic taurine release in rat cortical cup model (Estevez et al. 1999), the osmo-sensitive taurine release in cerebellar granule neurons is only prevented after chronic downregulation of PKC (Morales-Mulia et al. 2001). Thus, lipid-derived messengers are certainly involved in the cellular response to osmotic challenges. The complexity of the networks they are part of precludes, however, prediction of the outcome of the experiments aimed at elucidating their physiological role. Apparently, various cells have developed different responses; the reason for this phenomenon may lie in their differentiation into functionally diverse entities.

Cyclic nucleotides

The intracellular amounts of cAMP, as well as cAMP-dependent phosphorylation, have been demonstrated to be unaffected by osmotic cell shrinkage in duck and turkey erythrocytes (Kregenow et al. 1976; Alper et al. 1980). Initiation of the cAMP pathway is more likely to inhibit RVI, since cAMP-dependent activation of PKA can prevent the uptake of the organic osmolytes betaine and myo-inositol, as shown for MDCK cells (Preston et al. 1995). The hypertonic increase of intracellular cAMP suppresses the activation of Erk and p38; both MAP-kinases are known to be activated after hypertonic cell shrinkage in various systems (Orlic et al. 2002). However, cAMP is able to evoke cell swelling in sweat glands (Ohtsuyama et al. 1993). Osmotic cell swelling has been found to increase intracellular cAMP levels in various systems and cAMP leads to cell volume decrease (reviewed in Lang et al. 1998a).

In rectal gland tubular cells from *Squalus acanthias*, cAMP-mediated cell shrinkage occurs by stimulation of NaCl excretion. A PKA-dependent phosphorylation of CFTR-type Cl^- channels is followed by activation of Na^+-K^+-$2Cl^-$ cotransporter (Greger 1996). In the human hematopoietic myeloblastic leukemia cell line ML-1, outward-rectified Cl^- channels are activated by PKA (Xu and Lu 1994). The activation of volume-sensitive Cl^- currents is inhibited by cAMP in myocardial cells, indicating that activation of these channels requires dephosphorylation of a cAMP-dependent phosphorylation site (Hall et al. 1995; Nagasaki et al. 2000).

Although hypotonicity increases intracellular cAMP levels in S49 mouse lymphoma cells, cAMP is not involved in triggering volume regulatory processes (Watson et al. 1991). As shown for the hypotonic myo-inositol efflux in human NT2-N neurons, PKA does not contribute to the osmotic regulation of this transporter (Novak et al. 2000). The efflux of organic osmolytes can be inhibited by PKA, since inhibition of PKA enhances the speed of the hypotonic activation of anionic amino acid conductance in *Leishmania major* promastigotes (Vieira et al. 1997).

Intracellular calcium

An involvement of calcium in volume regulation has been found in many cellular systems. In particular, the regulation of cell swelling seems to often be accompanied by calcium signals. In the case of hypertonic stress and cell shrinkage, other intracellular signal transduction processes seem to be responsible for the recovery of the cell volume and thus the reports about an involvement of calcium in RVI are very rare (Marchenko and Sage 2000; Erickson et al. 2001). In this overview, we will focus, therefore, on the role of calcium during RVD.

The core process of volume regulation after cell swelling is a release of osmolytes, resulting in reduction of cellular water content. Many different transport processes involved in RVD have been found to be calcium-dependent. Calcium-regulated potassium conductance via BK_{Ca}, IK_{Ca}, and SK_{Ca} have been found in many cells (see the section entitled "K^+ channels"). Also chloride conductances induced by volume changes have been observed to be regulated by calcium/calmodulin-dependent processes (see the section entitled "Cl^- channels"). In guinea pig jejunal villus epithelial cells, a calcium/calmodulin kinase II-mediated phosphorylation is a critical determinant of the volume regulation (MacLeod and Hamilton 1999a). In some cells, the calcium action on the same transport system seems to be transmitted via different effector proteins; for example, in human cervical cancer cells, myosin light-chain kinase and protein kinase C regulate a volume-sensitive chloride channel (Chou et al. 1998; Shen et al. 2002). In astrocytes, chloride conductance as well as taurine release are processes regulated by calcium and calmodulin (Li et al. 2002). In rat IMCD cells, calcium is involved in the release of sorbitol (Ruhfus et al. 1996). Recently, a coupling between arginine vasopressin-induced intracellular calcium mobilization and apical exocytosis was investigated in isolated perfused rat IMCD segments (Yip 2002). Vasopressin induced a rapid increase of the intracellular calcium followed by sustained calcium oscillations and changes in cell volume. Apical exocytosis accompanied by an increase of osmotic water permeability could be prevented by an inhibition of intracellular calcium release. The study shows that in IMCD cells vasopressin, via

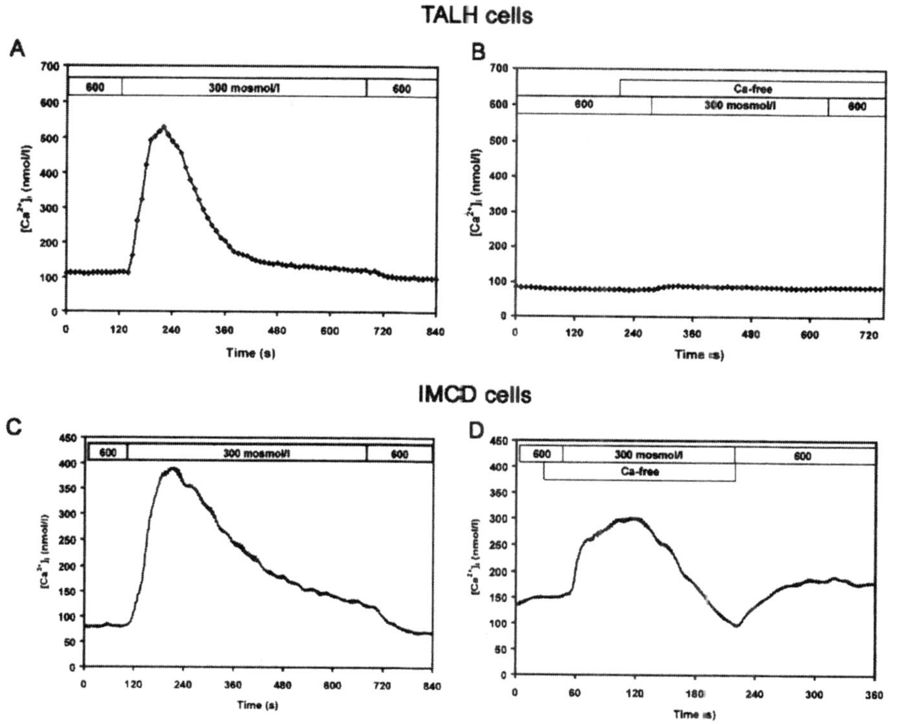

Fig. 9 Changes in the intracellular calcium concentration of TALH cells (**A**, **B**) and IMCD cells (**C**, **D**). At time indicated osmolarity was reduced from 600 mOsm/l to 300 mOsm/l by omission of sucrose. The experiments in calcium-containing solutions show a sustained calcium increase in both TALH (**A**) and IMCD (**C**) cells. In low-calcium solutions there was a calcium increase in the IMCD-cells due to a calcium release from intracellular stores (**D**). In TALH cells (**B**) no changes of the calcium concentration could be observed in calcium-free solutions as a prevention of the calcium influx made the calcium-induced calcium release impossible (**A** and **B** from Tinel et al. 2002; **C** and **D** from Tinel et al. 1994)

the V_2 receptor, triggers calcium oscillations which are required for exocytotic insertion of aquaporin-2 (Yip 2002). Activation of the sorbitol channel in rabbit TALH cells involves phosphorylation by a calcium/calmodulin kinase (Kinne-Saffran and Kinne 1997).

The source of the calcium involved in the volume regulation seems to be cell type-dependent. In several cells, RVD strictly depends on calcium entry across the plasma membrane. In these systems, the volume response can be inhibited by extracellular calcium removal or by addition of calcium channel blockers. A swelling-induced rise in intracellular calcium in rabbit medullary thick ascending limb cells was absent in low-calcium solutions (Montrose-Rafizadeh et al. 1991; see Fig. 9).

Similarly, strict dependence on calcium entry was found in, e.g.. osteosarcoma cells (Yamaguchi et al. 1989), rabbit proximal tubule cells (Suzuki et al. 1990), intestine 407 cells (Hazama and Okada 1988), and cultured mouse primary sensory neurons (Viana et al. 2001). The diversity of calcium channels which are responsible for the calcium influx during hypotonic stress in different cell types is very high. Cell swelling induces stretch of the plasma membrane; thus, it is possible that stretch-activated calcium channels are activated by this process. Indeed, there are many reports showing that Gd^{3+}, a nonspecific inhibitor of stretch-activated channels, can block the calcium increase during hypotonic stress (Chen et al. 1996b; Urbach et al. 1999; Miyauchi et al. 2000; Viana et al. 2001).

Stretch-activated calcium influx might, in some cells, be mediated by voltage-sensitive calcium pathways, as it has been shown in osteocytes (Miyauchi et al. 2000). In these cells, antisense oligodeoxynucleotides against the α 1C subunit of voltage-operated L-type calcium channels abolished the calcium increase during hypotonic stress. The expression of the α 1C subunit of voltage-operated L-type calcium channels has also been shown in UMR-106 cells (Kizer et al. 1999). The existence of a voltage-activated calcium L-type channel which may participate in the volume regulation has also been identified in the apical membrane of cultured rabbit proximal tubule cells using the patch-clamp technique (Zhang and O'Neil 1996). In many other cell types the calcium increase during hypotonic stress was sensitive to blockers of voltage-activated calcium channels (Montrose-Rafizadeh and Guggino 1991; Bender et al. 1994; Mignen et al. 1999). In inner medullary collecting duct cells (IMCD), diltiazem, nifedipine, and verapamil partly reduced the calcium increase (Tinel et al. 2000). However, the concentration of the blocker was quite high (40 μMol/l). Thus, calcium influx seems to occur via calcium channels which possess pharmacological characteristics different from those of classical voltage-gated channel. The fact that an inhibitor of receptor-mediated channels (SKF 96365) also reduced the calcium influx in IMCD cells supports the idea (Tinel et al. 2000). Channels involved in hypotonic calcium influx in thick ascending limb of Henle's loop (TALH) cells could be inhibited by nifedipine, but not diltiazem or verapamil (Tinel et al. 2002). These results can be correlated with the existence of homologous genes for different types of subunits and several types of auxiliary subunits of voltage-dependent calcium channels (Isom et al. 1994). It can also explain the difference compared to the characteristics of the calcium influx observed in TALH cells, which were not adapted to higher osmolarity (Montrose-Rafizadeh and Guggino 1991). In these cells, hypotonicity-induced calcium entry occurred via a nifedipine- and verapamil-sensitive pathway.

A complete inhibition of the calcium increase by calcium channel blockers or lack of the calcium signal in calcium-free solution has often been interpreted as proof that the intracellular calcium stores are not involved in the calcium rise during hypotonic stress. The assumption would be correct only if no ryanodine-sensitive calcium stores exist in the cell. Using caffeine, it has been shown that, e.g., TALH cells possess calcium stores that, although at first sight not involved in the volume regulation, can be mobilized experimentally in isotonic solution (Tinel et al. 2002). Caffeine causes calcium release from ryanodine-sensitive stores. The ryanodine receptor in the membrane of these stores is sensitive to calcium, which in micromolar concentration stimulates the receptor and triggers a calcium release (Tinel et al. 2000). This phenomenon is referred to as calcium-induced calcium release (CICR), a mechanism responsible for, e.g., heart muscle contraction. In TALH cells, the existence of CICR during hypotonic stimulation has been shown using manganese quenching experiments (see Figs. 9, 10).

Calcium release from intracellular stores followed the calcium influx during hypotonic stress (Tinel et al. 2002). The involvement of CICR in volume regulation has already been described in a human intestinal epithelial cell line (Intestine 407; Hazama and Okada 1990). For the volume regulation of these cells, the calcium influx is necessary, but not sufficient, unless it elicits CICR. In rabbit corneal epithelial cell, hypotonic stress has been found to induce a calcium release from ryanodine-sensitive stores followed by activation of a calcium influx (Wu et al. 1997). The authors describe the mechanism as CICR; how-

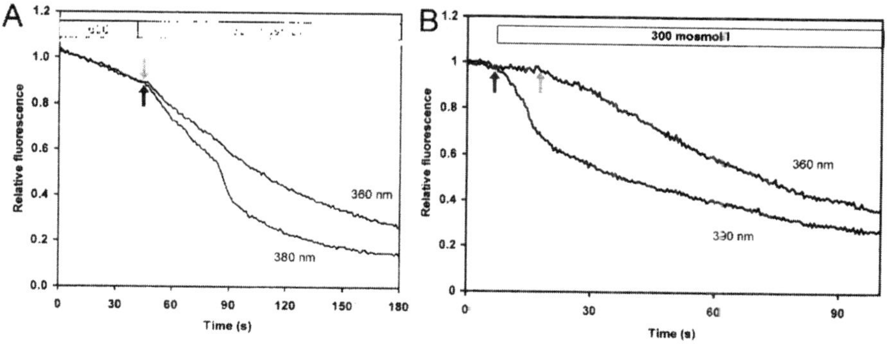

Fig. 10 Quench of fura-2 fluorescence by manganese under hypotonic conditions. The osmolarity was reduced from 600 mOsm/l to 300 mOsm/l by omission of sucrose. **A** In TALH cells, the hypotonic stress in calcium-containing solutions induced a monoexponential decay of the fluorescence intensity at 360 nm, reflecting a constant calcium influx. At 380 nm, the fluorescence decay was biphasic. During the first 40 s, both manganese and calcium entered the cell and induced a reduction of the fluorescence intensity. The additional fast decay was due to calcium-induced calcium release. **B** In IMCD cells hypotonic stress in calcium-free solutions induced a fast decay of the fluorescence at 390 nm due to calcium release from intracellular stores followed by an opening of a calcium influx pathway visible as a decrease in the fluorescence at 360 nm. → start of increase in intracellular calcium, ⇒start of calcium-entry from the extracellular medium (**A** from Tinel et al. 2002; **B** from Tinel et al. 1994)

ever, it corresponds more to the store-operated calcium influx. The RVD in these cells was also strictly dependent on the calcium rise. In TALH cells, the volume regulation was only partly calcium-dependent; thus, the calcium release from intracellular stores probably works as an amplifier of the signal. CICR is very important for the generation of calcium oscillations and calcium waves (Tinel et al. 2000). In some cells, there are initiation sites where the signal appears first before spreading through the cell as a wave (Tinel et al. 1999). Such local calcium events during the volume regulation of TALH cells could control membrane recycling, as has been described to occur in IMCD cells (Czekay et al. 1994). During exposure of IMCD cells to hypotonicity, the submembranal actin web rapidly disintegrates, and reserve vesicles, probably containing a sorbitol transporter, move to and fuse with the basolateral plasma membrane. The fusion causes a rapid increase in sorbitol permeability (see Fig. 4 above). These membrane areas are recovered by internalization, and the transport systems for sorbitol are concomitantly retrieved (Czekay et al. 1994).

The majority of studies concerning the role of calcium in volume regulation describe a biphasic calcium rise consisting of a calcium influx followed by calcium release from intracellular stores which are not ryanodine-sensitive. Manoeuvres leading to depletion of the intracellular stores or inhibiting the calcium release abolish the calcium increase in these cells during hypotonic stress (Hoffmann et al. 1984; McCarty and O'Neil 1991; Bender et al. 1994; Civan et al. 1994a; Mignen et al. 1999). Intracellular stores seem to participate in the shaping of the calcium signal when the changes of the cell volume occur in isotonic solution, e.g., in Ehrlich ascites tumor cells after stimulation with lysophosphatidic acid (Pedersen et al. 2000), or in pneumocytes after pretreatment with insulin (Marunaka et al. 1999). In some of the reports, the calcium stores are described as being inositol-trisphosphate (IP_3)-sensitive. An increase of the intracellular concentration of IP_3 has been measured during isotonic swelling of hepatocytes (Baquet et al. 1991), and during volume regulation of rabbit proximal tubule cells (Suzuki et al. 1990) and astrocytes (Bender et al. 1993).

Although in IMCD cells the calcium rise could be blocked by inhibitors of the calcium release from IP_3-sensitive stores, the signal transduction pathways leading to the calcium mobilization seem to be different (Tinel et al. 1994). In these cells, cell swelling in hypotonic solution is paralleled by an increase in intracellular calcium concentration typically appearing in two distinct phases. Whereas the first phase of the calcium response reflects predominantly calcium release from intracellular stores, the second sustained component is due to an increase in calcium entry from the extracellular compartment (Mooren and Kinne 1994; Tinel et al. 1994; see Figs. 9 and 10). However, measurement of intracellular IP_3 revealed no significant changes in the concentration of this second messenger (Tinel et al. 1997). Instead, inhibition of the arachidonic acid metabolism led to a considerable reduction of calcium release, as well as RVD. Thus, in these cells the calcium mobilization during RVD may be initiated by arachidonic acid or by its metabolites. Under isotonic conditions, an application of arachidonic acid induces a considerable increase in the intracellular calcium concentration in IMCD cells, both in high and in low calcium solutions, indicating that IMCD cells posses arachidonic acid-sensitive calcium stores. Extracellular ATP induces in these cells an increase in the intracellular IP_3 concentration and calcium rise, which reveals the existence of functionally active IP_3-sensitive stores. Thus, IMCD-cells possess arachidonic acid-sensitive as well as IP_3-sensitive calcium stores. These stores may be physically distinct; alternatively, the arachidonic acid-sensitive stores may be a part of the IP_3-sensitive pool. This is not contradictory to the finding that stimulation of calcium release by IP_3 during volume regulation apparently does not occur in IMCD cells. Arachidonic acid could change the affinity of the IP_3-receptor for IP_3 so that the IP_3 concentration already present in the cytosol of nonstimulated cells might induce a calcium release. It is possible that cooperation of different calcium stores control the volume regulation of these cells.

A combination of two different calcium-releasing mechanisms in the regulation of the cell volume has also been found in astrocytes (Bender and Norenberg 1994). In these cells, intracellular calcium release from CICR-stores and from IP_3-sensitive intracellular calcium stores play a role in the mechanism of potassium release under hyposmotic conditions. In submandibular gland acinar cells, the calcium increase is a result of a calcium release from agonist-sensitive stores and from mitochondria (Park 2002). Mitochondria are known to participate in modulating calcium signals in other cells (Hoffmann et al. 1984), and it is possible that they fulfill this function also during volume regulation, but this question remains to be elucidated.

The signal transduction processes leading to generation of the second messengers and to the calcium release (IP_3 or arachidonic acid) involves probably the action of phospholipases. Inhibitors of the phospholipase A_2 impaired the volume regulation of, e.g., human neuroblastoma cells (Basavappa et al. 1998) and human platelets (Margalit et al. 1993a). In cultured rabbit principal cells, apical application of the phospholipase A_2 agonist increased single-channel open probability of a volume-sensitive potassium channel (Ling et al. 1992). In IMCD cells, the activation of the arachidonic acid cascade during RVD occurs via activation of phospholipase C and probably phospholipase A_2 (Mooren and Kinne 1994; see also the section entitled “Arachidonic acid and eicosanoids”).

Usually phospholipases are activated by G-proteins. In IMCD cells, G-proteins are involved in the generation of the calcium signal (Mooren and Kinne 1994). Involvement of

G-proteins in the regulation of RVD has also been reported to occur in human platelets (Margalit et al. 1993a), and inhibition of RVD by pertussis toxin has been described in proximal tubule cells (Suzuki et al. 1990).

How G-proteins are actually activated during osmotic swelling remains obscure. A mechanoreceptor which could serve as a volume sensor has only been described for neurons (Oliet and Bourque 1993). A mechanical-biochemical coupling involving G-proteins and phospholipase A_2 during cell swelling has been postulated for platelets (Margalit et al. 1993a). OTRPC4, a nonselective calcium-permeable channel which belongs to the TRP channel family, has been proposed to be a candidate for a molecular sensor that confers osmosensitivity in mammalian cells (Strotmann et al. 2000). It exhibits spontaneous activity in isotonic media and is rapidly activated by a decrease (and is inhibited by an increase) of the extracellular osmolarity. Changes in osmolarity of as little as 10% result in significant changes in intracellular calcium concentration in HEK293 cells expressing the channel. This increase resulted exclusively from calcium influx from the extracellular space, with no involvement of calcium release from intracellular stores. In contrast to the members of the classical TRP channel family that are activated secondary to the activation of phospholipase C-coupled receptors, the OTRPCs have been proposed to be involved in the transduction of physical and chemical stimuli. However, OTRPC4 activity was unaffected by pressure-induced stretching of the membrane. The authors propose the channel as a universal sensor in mammalian cells, but the results from other studies where calcium signals could be abolished by inhibition of different steps of the signal transduction contradict this assumption.

In some cells, an autocrine (or paracrine) action of ATP has been found to initiate the calcium signalling during the volume regulation. During osmotic swelling, intracellular ATP is released to the extracellular space in a number of cells. In the immediate vicinity of the cell surface, released ATP has been shown to reach a concentration high enough to simulate P_2-purinergic receptors in a human epithelial cell line, intestine 407 (Dezaki et al. 2000). RVD in these cells is facilitated by ATP by augmenting intracellular calcium rise via the stimulation of purinergic P_{2Y2} receptors. Similar observations were made in endothelial cells from the rat caudal artery (Shinozuka et al. 2001). In hepatocytes, the signalling pathways seem to be more complicated. Ferenchak et al. have found a constitutive and volume-dependent ATP release, which was a critical determinant of membrane chloride permeability. P_2 receptor antagonist suramin abolished the volume recovery after exposure of hepatocytes to hypotonic buffer (Feranchak et al. 2000). However, Roe et al. have shown that the swelling-induced calcium signal in hepatocytes was unaffected by either extracellular ATP depletion or blockade of P_2 receptors (Roe et al. 2001). The swelling elicited an increase of the intracellular calcium, which was essential for ion channel activation and volume recovery, but this increase did not stem from activation of volume-sensitive P_2 receptors. Taken together, the results from both groups indicate that in hepatocytes two different pathways initiate the calcium signalling during hypotonic stress: a purinergic-dependent and a purinergic-independent pathway.

The mechanisms guaranteeing the interplay between the calcium release and calcium influx are not very well understood. In IMCD cells, the calcium release from the stores is a prerequisite for the activation of the calcium influx (Tinel et al. 1994). The opening of the plasma membrane ion channels happens possibly to prolong the calcium response beyond the point where intracellular stores are depleted and/or to support the refilling of these stores after calcium depletion. If the calcium release takes place in calcium-free isotonic solutions, a readdition of calcium to the extracellular medium induces a sustained increase of the intracellular calcium concentration. This indicates that the IMCD cells possess a capacitative calcium influx mechanism. The activation of this pathway is not dependent on an increase of the intracellular calcium concentration. Thus, there has to be another signal activating the calcium channels in the plasma membrane. In IMCD cells, the cytoskeleton could forward the message from the empty stores to the plasma membrane as Mooren and Kinne have observed a complete inhibition of the hypotonic calcium increase by cytochalasin B (Mooren and Kinne 1994). The signal transduction from the stores to the plasma membrane takes some 17 s, as could be estimated using the manganese quenching technique (Tinel et al. 1994; see Fig. 10).

In proximal straight tubule, a temporary dependence of the RVD on calcium has been observed (McCarty and O'Neil 1991). There is a short period of time (about 60 s), during which extracellular calcium is required. Outside of this "calcium window," RVD would inactivate and could not be reactivated by subsequent addition of calcium. It was found that the calcium permeability did not inactivate over several minutes, indicating that the temporary dependence of RVD on extracellular calcium is not due to the transient activation of calcium entry pathway (McCarty and O'Neil 1991).

A very interesting new aspect of the interaction between calcium stores and channels has been found by Lemonnier et al. in human prostate cancer epithelial cells (Lemonnier et al. 2002). In these cells, although variations of both intracellular and extracellular calcium concentrations had no visible effects on a volume-regulated anion channel, calcium influx via store-operated channels strongly modulates it. The authors suggest that interaction between the volume-regulated anion channel and calcium occurs in the confined compartments at the inner surface of the membrane that are not accessible to changes of global intracellular calcium. These domains can be readily reached by calcium entering the cell via plasma membrane, especially through the store-operated calcium channels. This preferred access of store-operated calcium channels to the volume-regulated anion channels suggests colocalization of these channels in the cell membrane (Lemonnier et al. 2002).

This finding belongs to the rare results illustrating the importance of the spatio-temporal aspect of calcium signalling. The intensive work during the last few years showed that, e.g., in secretory cells, the global calcium rise observed in many studies is mainly due to experimental manipulations and that local calcium events (sparks, oscillations, waves) reflect what actually happens under physiological conditions (Tinel et al. 1999; Tinel et al 2000). Future work concerning calcium signalling during volume regulation should concentrate much more on the local calcium events. It would be very important to recognise the mechanisms confining the signals in time and space to allow regulation of specific volume-related processes without influencing other cellular processes influenced by calcium.

Conclusions and perspectives

The data presented above document once more the complexity of events that ultimately result in the maintenance of cell volume and its restoration after disturbance by an osmotic difference between intra- and extracellular space. This complexity, although difficult to decipher, provides different cells with various mechanisms that are adapted to protect themselves and at the same time to fulfill their overall role in the organism. Within one particular cell, the presence of multiple systems might open multiple possibilities to achieve a graded response to osmotic stimuli.

To obtain further insight into this complexity several steps at different levels of resolution are necessary. At the level of the single molecules, proper molecular and pharmacological identification, proper biophysical characterization, and proper analysis of biochemical alterations have to be continued and intensified. At the cellular level, the simultaneous measurement of the activities of the most important osmolyte transporters and the concentration of the various osmolytes at a higher temporal and spatial resolution have to be performed. The same holds for the determination of the localization and activity of the various elements of the signal transduction network.

Only then will it be possible to describe the dynamic nature of cell volume regulation and its underlying mechanisms in a quantitative way based on causal relationships.

Acknowledgements. We thank Leon Goldstein, Rene Notais, Bernd Nilius and Markus Paulmichl for providing us with unpublished results of their current work. Stefan H. Boese was kind enough to comment on the chapters relating to chloride channels.

The expert and speedy secretarial help of Petra Jurczynski, the outstanding editorial assistance of Natascha Kist, the knowledgeable advice of Rosemarie Hübner and the professional creation of the artwork by Gesine Schulte are gratefully acknowledged.

Dedication: This chapter is dedicated to the many successful years of collaboration of the authors and to the excellent Ph.D. students and Postdocs, who shared our excitement.

References

Adelman JP, Shen KZ, Kavanaugh MP, Warren RA, Wu YN, Lagrutta A, Bond CT North RA (1992) Calcium-activated potassium channels expressed from cloned complementary DNAs. Neuron 9:209–216

Aharonovitz O, Granot Y (1996) Stimulation of mitogen-activated protein kinase and Na^+/H^+ exchanger in human platelets. Differential effect of phorbol ester and vasopressin. J Biol Chem 271:16494–16499

Al Habori M (2001) Macromolecular crowding and its role as intracellular signalling of cell volume regulation. Int J Biochem Cell Biol 33:844–864

Alessi DR, Deak M, Casamayor A, Caudwell FB, Morrice N, Norman DG, Gaffrey P, Reese CB, MacDougall CN, Harbison D, Ashworth A, Bownes M (1997) 3-Phosphoinositide-dependent protein kinase-1 (PDK1): structural and functional homology with the Drosophila DSTPK61 kinase. Curr Biol 7:776–789

Alper SL, Palfrey HC, DeRiemer SA, Greengard P (1980) Hormonal control of protein phosphorylation in turkey erythrocytes. Phosphorylation by cAMP-dependent and Ca^{2+}-dependent protein kinases of distinct sites in goblin, a high molecular weight protein of the plasma membrane. J Biol Chem 255:11029–11039

Alvarez de la Rosa D, Zhang P, Naray-Fejes-Toth A. Fejes-Toth G, Canessa CM (1999) The serum and glucocorticoid kinase sgk increases the abundance of epithelial sodium channels in the plasma membrane of Xenopus oocytes. J Biol Chem 274:37834–37839

Amano M, Chihara K, Kimura K, Fukata Y, Nakamura N, Matsuura Y, Kaibuchi K (1997) Formation of actin stress fibers and focal adhesions enhanced by Rho-kinase. Science 275:1308–1311

Amano M, Ito M, Kimura K, Fukata Y, Chihara K, Nakano T, Matsuura Y, Kaibuchi K (1996) Phosphorylation and activation of myosin by Rho-associated kinase (Rho- kinase) J Biol Chem 271:20246–20249

Atkinson NS, Robertson GA, Ganetzky B (1991) A component of calcium-activated potassium channels encoded by the Drosophila slo locus. Science 253 551–555

Atta MG, Dahl SC, Kwon HM, Handler JS (1999) Tyrosine kinase inhibitors and immunosuppressants perturb the myo-inositol but not the betaine cotransporter in isotonic and hypertonic MDCK cells. Kidney Int 55:956–962

Awayda MS, Subramanyam M (1998) Regulation of the epithelial Na^+ channel by membrane tension. J Gen Physiol 112:97–111

Bagnasco S, Balaban R, Fales HM, Yang YM, Burg M (1986) Predominant osmotically active organic solutes in rat and rabbit renal medullas. J Biol Chem 261:5872–5877

Bagnasco SM, Murphy HR, Bedford JJ, Burg MB (1988) Osmoregulation by slow changes in aldose reductase and rapid changes in sorbitol flux. Am J Physiol 254:C788–C792

Bai L, Collins JF, Muller YL, Xu H, Kiela PR, Ghishan FK (1999) Characterization of cis-elements required for osmotic response of rat Na^+/H^+ exchanger-2 (NHE-2) gene. Am J Physiol 277:R1112–R1119

Banderali U, Roy G (1992a) Activation of K^+ and Cl^- channels in MDCK cells during volume regulation in hypotonic media. J Membr Biol 126:219–234

Banderali U, Roy G (1992b) Anion channels for amino acids in MDCK cells. Am J Physiol 263:C1200–C1207.

Baquet A, Meijer AJ, Hue L (1991) Hepatocyte swelling increases inositol 1,4,5-trisphosphate, calcium and cyclic AMP concentration but antagonizes phosphorylase activation by Ca^{2+}-dependent hormones. FEBS Lett 278:103–106

Barhanin J, Lesage F, Guillemare E, Fink M, Lazdunski M, Romey G (1996) KvLQT1 and IsK (minK) proteins associate to form the IKs cardiac potassium current. Nature 384:78–80

Basavappa S, Pedersen SF, Jorgensen NK, Ellory JC, Hoffmann EK (1998) Swelling-induced arachidonic acid release via the 85-kDa cPLA2 in human neuroblastoma cells. J Neurophysiol 79:1441–1449

Bauernschmitt HG, Kinne RKH (1993) Metabolism of the "organic osmolyte" glycerophosphorylcholine in isolated rat inner medullary collecting duct cells. II. Regulation by extracellular osmolality. Biochim Biophys Acta 1150:25–34

Bear CE (1990) A nonselective cation channel in rat liver cells is activated by membrane stretch. Am J Physiol 258:C421–C428

Bear CE (1991) A K^+-selective channel in the colonic carcinoma cell line: CaCo-2 is activated with membrane stretch. Biochim Biophys Acta 1069:267–272

Beck F, Dörge A, Rick R, Thurau K (1985) Osmoregulation of renal papillary cells. Pflugers Arch 405 Suppl 1, S28–S32

Beck FX, Burger-Kentischer A, Müller E (1998) Cellular response to osmotic stress in the renal medulla. Pflugers Arch 436:814–827

Bedford JJ, Bagnasco SM, Kador PF, Harris HW Jr., Burg MB (1987) Characterization and purification of a mammalian osmoregulatory protein, aldose reductase, induced in renal medullary cells by high extracellular NaCl. J Biol Chem 262:14255–14259

Beetsch JW, Olson JE (1996) Hyperosmotic exposure alters total taurine quantity and cellular transport in rat astrocyte cultures. Biochim Biophys Acta 1290:141–148

Behrens R, Nolting A, Reimann F, Schwarz M, Waldschütz R, Pongs O (2002) hKCNMB3 and hKCNMB4, cloning and characterization of two members of the large-conductance calcium-activated potassium β subunit family. FEBS Lett 474:99–106

Bender AS, Norenberg MD (1994) Calcium dependence of hypo-osmotically induced potassium release in cultured astrocytes. J Neurosci 14:4237–4243

Bender AS, Mantelle LL, Norenberg MD (1994) Stimulation of calcium uptake in cultured astrocytes by hypo-osmotic stress-effect of cyclic AMP. Brain Res 645:27–35

Bender AS, Neary JT, Norenberg MD (1993) Role of phosphoinositide hydrolysis in astrocyte volume regulation. J Neurochem 61:1506–1514

Benos DJ, Fuller CM, Shlyonsky V Gh, Berdiev BK, Ismailov II (1997) Amiloride-sensitive Na^+ channels: Insights and outlooks. News Physiol Sci 12:55–61

Bertrand B, Wakabayashi S, Ikeda T, Pouysségur J, Shigekawa M (1994) The Na^+/H^+ exchanger isoform 1 (NHE1) is a novel member of the calmodulin-binding proteins. J Biol Chem 269:13703–13709

Bianchini L, L'Allemain G, Pouyssegur J (1997) The p42/p44 mitogen-activated protein kinase cascade is determinant in mediating activation of the Na^+/H^+ exchanger (NHE1 isoform) in response to growth factors. J Biol Chem 272:271–279

Blanco G, Mercer RW (1998) Isozymes of the Na-K-ATPase: heterogeneity in structure, diversity in function. Am J Physiol 275:F633–F650

Bleich M, Riedemann N, Warth R, Kerstan D, Leipziger J, Hör M, van Driessche W, Greger R (1996) Ca^{2+} regulated K^+ and nonselective cation channels in the basolateral membrane of rat colonic crypt base cells. Pflugers Arch 432:1011–1022

Boese SH, Glanville M, Gray MA, Simmons NL (2000) The swelling-activated anion conductance in the mouse renal inner medullary collecting duct cell line mIMCD-K2. J Membr Biol 177:51–64

Boese SH, Kinne RKH, Wehner F (1996a) Single-channel properties of swelling-activated anion conductance in rat inner medullary collecting duct cells. Am J Physiol 271:F1224–F1233

Boese SH, Wehner F, Kinne RKH (1996b) Taurine permeation through swelling-activated anion conductance in rat IMCD cells in primary culture. Am J Physiol 271:F498–F507

Böhmer C, Wagner CA, Beck S, Moschen I, Melzig J, Werner A, Lin J-T, Lang F, Wehner F (2000) The shrinkage-activated Na^+ conductance of rat hepatocytes and its possible correlation to rENaC. Cell Physiol Biochem 10:187–194

Böhmer C, Wehner F (2001) The epithelial Na^+ channel (ENaC) is related to the hypertonicity-induced Na^+ conductance in rat hepatocytes. FEBS Lett 494:125–128

Bookstein C, Musch MW, DePaoli A, Xie Y, Villereal M, Rao MC, Chang EB (1994) A unique sodium-hydrogen exchange isoform (NHE-4) of the inner medulla of the rat kidney is induced by hyperosmolarity. J Biol Chem 269:29704-29709

Bookstein C, Xie Y, Rabenau K, Musch MW, McSwine RL, Rao MC, Chang EB (1997) Tissue distribution of Na^+/H^+ exchanger isoforms NHE2 and NHE4 in rat intestine and kidney. Am J Physiol 273:C1496–C1505

Boron WF (2001) Sodium-coupled bicarbonate transporters. JOP 2:176–181

Bowen JW (1992) Regulation of Na^+-K^+-ATPase expression in cultured renal cells by incubation in hypertonic medium. Am J Physiol 262:C845–C853

Brandsch M, Miyamoto Y, Ganapathy V, Leibach FH (1993) Regulation of taurine transport in human colon carcinoma cell lines (HT- 29 and Caco-2) by protein kinase C. Am J Physiol 264:G939–G946

Brierley GP, Davis MH, Cragoe EJ Jr., Jung DW (1989) Kinetic properties of the Na^+/H^+ antiport of heart mitochondria. Biochemistry (Mosc) 28:4347–4354

Brosius FC III, Alper SL, Garcia AM, Lodish HF (1989) The major kidney band 3 gene transcript predicts an amino-terminal truncated band 3 polypeptide. J Biol Chem 264:7784–7787

Burg MB (1995) Molecular basis of osmotic regulation. Am J Physiol 268:F983–F996

Burg MB (1996) Coordinate regulation of organic osmolytes in renal cells. Kidney Int 49:1684–1685

Burg MB, Kador PF (1988) Sorbitol, osmoregulation, and the complications of diabetes. J Clin Invest 81:635–640

Burg MB, Kwon ED, Kültz D (1997) Regulation of gene expression by hypertonicity. Annu Rev Physiol 59:437–455

Burger-Kentischer A, Müller E, Neuhofer W, März J, Thurau K, Beck F (1999) Expression of aldose reductase, sorbitol dehydrogenase and Na^+/myo- inositol and Na^+/Cl^-/betaine transporter mRNAs in individual cells of the kidney during changes in the diuretic state. Pflugers Arch 437:248–254

Burridge K, Chrzanowska-Wodnicka M (1996) Focal adhesions, contractility, and signaling. Annu Rev Cell Dev Biol 12:463–518

Burridge K, Zhong C (1997) Focal adhesion assembly. Trends Cell Biol 7:342–347

Busch AE, Maylie J (1993) MinK channels: A minimal channel protein with a maximal impact. Cell Physiol Biochem 3:270–276

Busch AE, Suessbrich H (1997) Role of the ISK protein in the IminK channel complex. Trends Pharmacol Sci 18:26–29

Busch AE, Varnum M, Adelman JP, North RA (1992) Hypotonic solution increases the slowly activating potassium current ISK expressed in *Xenopus* oocytes. Biochem Biophys Res Commun 184:804–810

Buyse G, Voets T, Tytgat J, De Greef C, Droogmans G, Nilius B, Eggermont J (1997) Expression of human pI_{Cln} and ClC-6 in Xenopus oocytes induces an identical endogenous chloride conductance. J Biol Chem 272:3615–3621

Cammarata PR, Schafer G, Chen SW, Guo Z, Reeves RE (2002) Osmoregulatory alterations in taurine uptake by cultured human and bovine lens epithelial cells. Invest Ophthalmol Vis Sci 43:425–433

Capasso JM, Rivard CJ, Berl T (2001) Long-term adaptation of renal cells to hypertonicity: role of MAP kinases and Na-K-ATPase. Am J Physiol Renal Physiol 280, F768-F776

Caruso-Neves C, Lopes AG (2000) Sodium pumps in the Malpighian tubule of Rhodnius sp. An Acad Bras Cienc 72:407–412

Chan HC, Fu WO, Chung YW, Huang SJ, Chan PSF, Wong PYD (1994) Swelling-induced anion and cation conductances in human epididymal cells. J Physiol 478:449–460

Chan HC, Nelson DJ (1992) Chloride-dependent cation conductance activated during cellular shrinkage. Science 257:669–671

Chen JG, Coe M, McAteer JA, Kempson SA (1996a) Hypertonic activation and recovery of system A amino acid transport in renal MDCK cells. Am J Physiol 270:F419–F424

Chen Y, Simasko SM, Niggel J, Sigurdson WJ, Sachs F (1996b) Ca^{2+} uptake in GH3 cells during hypotonic swelling: the sensory role of stretch-activated ion channels. Am J Physiol 270:C1790–C1798

Chen L, Wang L, Jacob TJ (1999a) Association of intrinsic pICln with volume-activated Cl^- current and volume regulation in a native epithelial cell. Am J Physiol 276:C182–C192

Chen SY, Bhargava A, Mastroberardino L, Meijer OC, Wang J, Buse P, Firestone GL, Verrey F, Pearce D (1999b) Epithelial sodium channel regulated by aldosterone-induced protein sgk. Proc Natl Acad Sci USA 96:2514–2519

Chesney RW, Gusowski N, Dabbagh S (1985a) Renal cortex taurine content regulates renal adaptive response to altered dietary intake of sulfur amino acids. J Clin Invest 76:2213–2221

Chesney RW, Gusowski N, Dabbagh S, Theissen M, Padilla M, Diehl A (1985b) Factors affecting the transport of β-amino acids in rat renal brush-border membrane vesicles. The role of external chloride. Biochim Biophys Acta 812:702–712

Chong LD, Traynor-Kaplan A, Bokoch GM, Schwartz MA (1994) The small GTP-binding protein Rho regulates a phosphatidylinositol 4-phosphate 5-kinase in mammalian cells. Cell 79:507–513

Chou CY, Shen MR, Hsu KS, Huang HY, Lin HC (1998) Involvement of PKC-α in regulatory volume decrease responses and activation of volume-sensitive chloride channels in human cervical cancer HT-3 cells. J Physiol 512:435–448

Chow CW, Woodside M, Demaurex N, Yu FH, Plant P, Rotin D, Grinstein S, Orlowski J (1999) Proline–rich motifs of the Na^+/H^+ exchanger 2 isoform. Binding of Src homology domain 3 and role in apical targeting in epithelia. J Biol Chem 274:10481–10488

Christensen O (1987) Mediation of cell volume regulation by Ca^{2+} influx through stretch-activated channels. Nature 330:66–68

Christensen O, Hoffmann EK (1992) Cell swelling activates K^+ and Cl^- channels as well as nonselective, stretch-activated cation channels in Ehrlich ascites tumor cells. J Membr Biol 129:13–36

Christensen O, Simon M, Randlev T (1989) Anion channels in a leaky epithelium. A patch-clamp study of choroid plexus. Pflugers Arch 415:37–46

Civan MM, Coca-Prados M, Peterson-Yantorno K (1994a) Pathways signaling the regulatory volume decrease of cultured nonpigmented ciliary epithelial cells. Invest Ophthalmol Vis Sci 35:2876–2886

Civan MM, Marano CW, Matschinsky FW, Peterson-Yantorno K (1994b) Prolonged incubation with elevated glucose inhibits the regulatory response to shrinkage of cultured human retinal pigment epithelial cells. J Membr Biol 139:1–13

Civan MM, Coca-Prados M, Peterson-Yantorno K (1996) Regulatory volume increase of human nonpigmented ciliary epithelial cells. Exp Eye Res 62:627–639

Civan MM, Peterson-Yantorno K, Sánchez-Torres J, Coca-Prados M (1997) Potential contribution of epithelial Na^+ channel to net secretion of aqueous humor. J Exp Zool 279:498–503

Clark EA, King WG, Brugge JS, Symons M, Hynes RO (1998) Integrin-mediated signals regulated by members of the rho family of GTPases. J Cell Biol 142:573–586

Clemo HF, Baumgarten CM (1997) Swelling-activated Gd^{3+}-sensitive cation current and cell volume regulation in rabbit ventricular myocytes. J Gen Physiol 110:297–312

Coca-Prados M, Sanchez-Torres J, Peterson-Yantorno K, Civan MM (1996) Association of ClC-3 channel with Cl^- transport by human nonpigmented ciliary epithelial cells. J Membr Biol 150:197–208

Cockcroft S (2001) Signalling roles of mammalian phospholipase D1 and D2. Cell Mol Life Sci 58:1674–1687

Cornet M, Ubl J, Kolb H-A (1993) Cytoskeleton and ion movements during volume regulation in cultured PC12 cells. J Membr Biol 133:161–170

Costa PMF, Fernandes PL, Ferreira HG, Ferreira KTG, Giraldez F (1987) Effects of cell volume changes on membrane ionic permeabilities and sodium transport in frog skin (*Rana ridibunda*). J Physiol (Lond) 393:1–17

Coulombe A, Coraboeuf E (1992) Large-conductance chloride channels of new-born rat cardiac myocytes are activated by hypotonic media. Pflugers Arch 422:143–150

Counillon L, Pouyssegur J (2000) The expanding family of eukaryotic Na^+/H^+ exchangers. J Biol Chem 275:1–4

Coupaye-Gerard B, Bookstein C, Duncan P, Chen XY, Smith PR, Musch M, Ernst SA, Chang EB, Kleyman TR (1996) Biosynthesis and cell surface delivery of the NHE1 isoform of Na^+/H^+ exchanger in A6 cells. Am J Physiol 271:C1639–C1645

Craelius W, Ross MJ, Harris DR, Chen VK, Palant CE (1993) Membrane currents controlled by physical forces in cultured mesangial cells. Kidney Int 43:535–543

Cunningham SA, Awayda MS, Bubien JK, Ismailov II, Arrate MP, Berdiev BK, Benos DJ, Fuller CM (1995) Cloning of an epithelial chloride channel from bovine trachea. J Biol Chem 270:31016–31026

Cuppoletti J, Tewari KP, Sherry AM, Kupert EY, Malinowska DH (2001) ClC-2 Cl- channels in human lung epithelia: activation by arachidonic acid, amidation, and acid-activated omeprazole. Am J Physiol Cell Physiol 281:C46–C54

Curtis DR, Johnston GA (1974) Amino acid transmitters in the mammalian central nervous system. Ergeb Physiol 69:97–188

Czekay RP, Kinne-Saffran E, Kinne RKH (1994) Membrane traffic and sorbitol release during osmo- and volume regulation in isolated rat renal inner medullary collecting duct cells. Eur J Cell Biol 63:20–31

de Angelis E, Petronini PG, Borghetti P, Borghetti AF, Wheeler KP (1999) Induction of betaine-gamma-aminobutyric acid transport activity in porcine chondrocytes exposed to hypertonicity. J Physiol 518 (Pt 1), 187–194

Dermietzel R, Hwang TK, Buettner R, Hofer A, Dotzler E, Kremer M, Deutzmann R, Thinnes FP, Fishman GI, Spray DC, (1994) Cloning and in situ localization of a brain-derived porin that constitutes a large-conductance anion channel in astrocytic plasma membranes. Proc Natl Acad Sci USA 91:499–503

Deutsch C, Chen L-Q (1993) Heterologous expression of specific K^+ channels in T lymphocytes: Functional consequences for volume regulation. Proc Natl Acad Sci USA 90:10036–10040

Devor DC, Frizzell RA (1998) Modulation of K^+ channels by arachidonic acid in T84 cells. II. Activation of a Ca^{2+}-independent K^+ channel. Am J Physiol 274:C149–C160

Dezaki K, Tsumura T, Maeno E, Okada Y (2000) Receptor-mediated facilitation of cell volume regulation by swelling-induced ATP release in human epithelial cells. Jpn J Physiol 50:235–241

Di Fulvio M, Lauf PK, Adragna NC (2001a) Nitric oxide signaling pathway regulates potassium chloride cotransporter-1 mRNA expression in vascular smooth muscle cells. J Biol Chem 276:44534–44540

Di Fulvio M, Lincoln TM, Lauf PK, Adragna NC (2001b) Protein kinase G regulates potassium chloride cotransporter-3 expression in primary cultures of rat vascular smooth muscle cells. J Biol Chem 276:21046–21052

Donaldson PJ, Chen LK, Lewis SA (1989) Effects of serosal anion composition on the permeability properties of rabbit urinary bladder. Am J Physiol 256:F1125–F1134

Doroshenko P, Neher E (1992) Volume-sensitive chloride conductance in bovine chromaffin cell membrane. J Physiol (Lond) 449:197–218

Duan D, Winter C, Cowley S, Hume JR, Horowitz B (1997) Molecular identification of a volume-regulated chloride channel. Nature 390:417–421

Duan D, Ye L, Britton F, Horowitz B, Hume JR (2000) A novel anionic inward rectifier in native cardiac myocytes. Circ Res 86:E63–E71

Dubé L, Parent L, Sauvé R (1990) Hypotonic shock activates a maxi K^+ channel in primary cultured proximal tubule cells. Am J Physiol 259:F348–F356

Dubinsky WP, Mayorga-Wark O, Schultz SG (2000) Potassium channels in basolateral membrane vesicles from necturus enterocytes: stretch and ATP sensitivity. Am J Physiol Cell Physiol 279:C634–C638

Dudeja PK, Rao DD, Syed I, Joshi V, Dahdal RY, Gardner C, Risk MC, Schmidt L, Bavishi D, Kim KE, Harig JM, Goldstein JL, Layden TJ, Ramaswamy K (1996) Intestinal distribution of human Na^+/H^+ exchanger isoforms NHE-1, NHE- 2, and NHE-3 mRNA. Am J Physiol 271:G483–G493

Dunbar LA, Caplan MJ (2000) The cell biology of ion pumps: sorting and regulation. Eur J Cell Biol 79:557–563

Dutzler R, Campbell EB, Cadene M, Chait BT, MacKinnon R (2002) X-ray structure of a ClC chloride channel at 3.0 A reveals the molecular basis of anion selectivity. Nature 415:287–294

Duzgun SA, Rasque H, Kito H, Azuma N, Li W, Basson MD, Gahtan V, Dudrick SJ, Sumpio BE (2000) Mitogen-activated protein phosphorylation in endothelial cells exposed to hyperosmolar conditions. J Cell Biochem 76:567–571

Ellis RJ (2001) Macromolecular crowding: obvious but underappreciated. Trends Biochem Sci 26:597–604

Emma F, McManus M, Strange K (1997) Intracellular electrolytes regulate the volume set point of the organic osmolyte/anion channel VSOAC. Am J Physiol 272:C1766–C1775

Erickson GR, Alexopoulos LG, Guilak F (2001) Hyperosmotic stress induces volume change and calcium transients in chondrocytes by transmembrane, phospholipid, and G-protein pathways. J Biomech 34:1527–1535

Estevez AY, O'Regan MH, Song D, Phillis JW (1999) Hypo-osmotically induced amino acid release from the rat cerebral cortex: role of phospholipases and protein kinases. Brain Res 844:1–9

Falke LC, Misler S (1989) Activity of ion channels during volume regulation by clonal N1E115 neuroblastoma cells. Proc Natl Acad Sci USA 86:3919–3923

Farrugia G, Rae J (1993) Effect of volume changes on a potassium current in rabbit corneal epithelial cells. Am J Physiol 264:C1238–C1245

Felipe A, Snyders DJ, Deal KK, Tamkun MM (1993) Influence of cloned voltage-gated K^+ channel expression on alanine transport, Rb^+ uptake, and cell volume. Am J Physiol 265:C1230–C1238

Feranchak AP, Berl T, Capasso J, Wojtaszek PA, Han J, Fitz JG (2001) p38 MAP kinase modulates liver cell volume through inhibition of membrane Na^+ permeability. J Clin Invest 108:1495–1504

Feranchak AP, Fitz JG, Roman RM (2000) Volume-sensitive purinergic signaling in human hepatocytes. J Hepatol 33:174–182

Feranchak AP, Roman RM, Doctor RB, Salter KD, Toker A, Fitz JG (1999) The lipid products of phosphoinositide 3-kinase contribute to regulation of cholangiocyte ATP and chloride transport. J Biol Chem 274:30979–30986

Feranchak AP, Roman RM, Schwiebert EM, Fitz JG (1998) Phosphatidylinositol 3-kinase contributes to cell volume regulation through effects on ATP release. J Biol Chem 273:14906–14911

Ferraris JD, Burg MB, Williams CK, Peters EM, Garcia-Perez A (1996) Betaine transporter cDNA cloning and effect of osmolytes on its mRNA induction. Am J Physiol 270:C650–C654

Fievet B, Gabillat N, Borgese F, Motais R (1995) Expression of band 3 anion exchanger induces chloride current and taurine transport: structure-function analysis. EMBO J 14:5158–5169

Fievet B, Perset F, Gabillat N, Guizouarn H, Borgese F, Ripoche P, Motais R (1998) Transport of uncharged organic solutes in Xenopus oocytes expressing red cell anion exchangers (AE1 s) Proc Natl Acad Sci USA 95:10996–11001

Filipovic D, Sackin H (1991) A calcium-permeable stretch-activated cation channel in renal proximal tubule. Am J Physiol 260:F119–F129

Filipovic D, Sackin H (1992) Stretch- and volume-activated channels in isolated proximal tubule cells. Am J Physiol 262:F857–F870

Forlani G, Bossi E, Perego C, Giovannardi S, Peres A (2001) Three kinds of currents in the canine betaine-GABA transporter BGT-1 expressed in *Xenopus laevis* oocytes. Biochim Biophys Acta 1538:172–180

Fraser CL, Swanson RA (1994) Female sex hormones inhibit volume regulation in rat brain astrocyte culture. Am J Physiol 267:C909–C914

Fuller CM, Benos DJ (2000) Ca^{2+}-activated Cl^- channels: a newly emerging anion transport family. News Physiol Sci 15:165–171

Fuller CM, Ji HL, Tousson AM, Elble RC, Pauli BU, Benos DJ (2001) Ca^{2+}-activated Cl^- channels: a newly emerging anion transport family. Pflugers Arch 443 Suppl 1, S107–S110

Furlong TJ, Moriyama T, Spring KR (1991) Activation of osmolyte efflux from cultured renal papillary epithelial cells. J Membr Biol 123:269–277

Fürst J, Bazzini C, Jakab M, Meyer G, König M, Gschwentner M, Ritter M, Schmarda A, Bottà G, Benz R, Deetjen P, Paulmichl M (2000a) Functional reconstitution of ICln in lipid bilayers. Pflugers Arch 440:100–115

Fürst J, Jakab M, Konig M, Ritter M, Gschwentner M, Rudzki J, Danzl J, Mayer M, Burtscher CM, Schirmer J, Maier B, Nairz M, Chwatal S, Paulmichl M (2000b) Structure and function of the ion channel ICln. Cell Physiol Biochem 10:329–334

Fyfe GK, Quinn A, Canessa CM (1998) Structure and function of the Mec-ENaC family of ion channels. Semin Nephrol 18:138–151

Gabbay KH (1973) The sorbitol pathway and the complications of diabetes. N Engl J Med 288:831–836

Gamper N, Huber SM, Badawi K, Lang F (2000) Cell volume-sensitive sodium channels upregulated by glucocorticoids in U937 macrophages. Pflugers Arch 441:281–286

Garavaglia L, Rodighiero S, Bertocchi C, Manfredi R, Fürst J, Gschwentner M, Ritter M, Bazzini C, Botta G, Jakab M, Meyer G, Paulmichl M (2002) ICln channels reconstituted in heart-lipid bilayer are selective to chloride. Pflugers Arch 443:748–753

Garcia-Calvo M, Knaus HG, McManus OB, Giangiacomo KM, Kaczorowski GJ, Garcia ML (1994) Purification and reconstitution of the high-conductance, calcium-activated potassium channel from tracheal smooth muscle. J Biol Chem 269:676–682

Garcia-Perez A, Burg MB (1991) Renal medullary organic osmolytes. Physiol Rev 71:1081–1115

Gardos G (2002) The function of calcium in the potassium permeability of human erythrocytes. Biochim Biophys Acta 30:653–654

Garty H, Furlong TJ, Ellis DE, Spring KR (1991) Sorbitol permease: an apical membrane transporter in cultured renal papillary epithelial cells. Am J Physiol 260:F650–F656

Garty H, Palmer LG (1997) Epithelial sodium channels: Function, structure, and regulation. Physiol Rev 77:359–396

Gill DR, Hyde SC, Higgins CF, Valverde MA, Mintenig GM, Sepúlveda FV (1992) Separation of drug transport and chloride channel functions of the human multidrug resistance P-glycoprotein. Cell 71:23–32

Gillen CM, Brill S, Payne JA, Forbush B, III (1996) Molecular cloning and functional expression of the K-Cl cotransporter from rabbit, rat, and human—A new member of the cation-chloride cotransporter family. J Biol Chem 271:16237–16244

Goetzl EJ, An S (1998) Diversity of cellular receptors and functions for the lysophospholipid growth factors lysophosphatidic acid and sphingosine 1-phosphate. FASEB J 12:1589–1598

Goldstein L, Davis EM (1994) Taurine, betaine, and inositol share a volume-sensitive transporter in skate erythrocyte cell membrane. Am J Physiol 267:R426–R431

Goldstein L, Davis-Amaral EM, Musch MW (1996) Organic osmolyte channels: transport characteristics and regulation. Kidney Int 49:1690–1694

Good DW, Di Mari JF, Watts BA, III (2000) Hypo-osmolality stimulates Na^+/H^+ exchange and HCO_3^- absorption in thick ascending limb via PI 3-kinase. Am J Physiol Cell Physiol 279:C1443–C1454

Gosling M, Poyner DR, Smith JW (1996) Effects of arachidonic acid upon the volume-sensitive chloride current in rat osteoblast-like (ROS 17/2.8) cells. J Physiol 493:613–623

Graf J, Haddad P, Häussinger D, Lang F (1988) Cell volume regulation in liver. Renal Physiol Biochem 11:202–220

Graf J, Häussinger D (1996) Ion transport in hepatocytes: Mechanisms and correlations to cell volume, hormone actions and metabolism. J Hepatol 24 Suppl. 1:53–77

Greger R (1996) The membrane transporters regulating epithelial NaCl secretion. Pflugers Arch 432:579–588

Grinstein S, Rothstein A, Cohen S (1985) Mechanism of osmotic activation of Na^+/H^+ exchange in rat thymic lymphocytes. J Gen Physiol 85:765–787

Grinstein S, Woodside M, Sardet C, Pouyssegur J, Rotin D (1992) Activation of the Na^+/H^+ antiporter during cell volume regulation. Evidence for a phosphorylation-independent mechanism. J Biol Chem 267:23823–23828

Grinstein S, Woodside M, Waddell TK, Downey GP, Orlowski J, Pouyssegur J, Wong DC, Foskett JK (1993) Focal localization of the NHE-1 isoform of the Na^+/H^+ antiport: assessment of effects on intracellular pH. EMBO J 12:5209–5218

Gründer S, Thiemann A, Pusch M, Jentsch TJ (1992) Regions involved in the opening of CIC-2 chloride channel by voltage and cell volume. Nature 360:759–762

Grunewald RW, Kinne RKH (1989) Intracellular sorbitol content in isolated rat inner medullary collecting duct cells. Regulation by extracellular osmolarity. Pflugers Arch 414:178–184

Grunewald RW, Kinne RKH (1999) Osmoregulation in the mammalian kidney: the role of organic osmolytes. J Exp Zool 283:708–724

Grunewald JM, Grunewald RW, Kinne RKH (1993a) Ion content and cell volume in isolated collecting duct cells: effect of hypotonicity. Kidney Int 44:509–517

Grunewald RW, Weber II, Kinne-Saffran E, Kinne RKH (1993b) Control of sorbitol metabolism in renal inner medulla of diabetic rats: regulation by substrate, cosubstrate and products of the aldose reductase reaction. Biochim Biophys Acta 1225:39–47

Grunewald JM, Grunewald RW, Kinne RKH (1994) Regulation of ion content and cell volume in isolated rat renal IMCD cells under hypertonic conditions. Am J Physiol 267:F13–F19

Grunewald RW, Weber II, Kinne RKH (1995) Renal inner medullary sorbitol metabolism. Am J Physiol 269:F696–F701

Grunewald RW, Wagner M, Schubert I, Franz HE, Müller GA, Steffgen J (1998) Rat renal expression of mRNA coding for aldose reductase and sorbitol dehydrogenase and its osmotic regulation in inner medullary collecting duct cells. Cell Physiol Biochem 8:293–303

Gschwentner M, Nagl UO, Woll E, Schmarda A, Ritter M, Paulmichl M (1995) Antisense oligonucleotides suppress cell-volume-induced activation of chloride channels. Pflugers Arch 430:464–470

Guastella J, Nelson N, Nelson H, Czyzyk L, Keynan S, Miedel MC, Davidson N, Lester HA, Kanner BI (1990) Cloning and expression of a rat brain GABA transporter. Science 249:1303–1306

Guizouarn H, Gabillat N, Motais R, Borgese F (2001) Multiple transport functions of a red blood cell anion exchanger, tAE1: its role in cell volume regulation. J Physiol (Lond) 535:497–506

Gusev GP, Sherstobitov AO (1996) An amiloride-sensitive, volume-dependent Na^+ transport across the Lamprey (*Lampetra fluviatilis*) erythrocyte membrane. Gen Physiol Biophys 15:129–143

Haas M, Forbush B, III (2000) The Na-K-Cl cotransporter of secretory epithelia. Annu Rev Physiol 62:515–534

Hafting T, Sand O (2000) Purinergic activation of BK channels in clonal kidney cells (Vero cells) Acta Physiol Scand 170:99–109

Hager K, Hazama A, Kwon HM, Loo DD, Handler JS, Wright EM (1995) Kinetics and specificity of the renal Na^+/myo-inositol cotransporter expressed in *Xenopus* oocytes. J Membr Biol 143:103–113

Hall AC, Bush PG (2001) The role of a swelling-activated taurine transport pathway in the regulation of articular chondrocyte volume. Pflugers Arch 442:771–781

Hall JA, Kirk J, Potts JR, Rae C, Kirk K (1996) Anion channel blockers inhibit swelling-activated anion, cation, and nonelectrolyte transport in HeLa cells. Am J Physiol 271:C579–C588

Hall SK, Zhang J, Lieberman M (1995) Cyclic AMP prevents activation of a swelling-induced chloride-sensitive conductance in chick heart cells. J Physiol 488:359–369

Hallows KR, Packman CH, Knauf PA (1991) Acute cell volume changes in anisotonic media affect F-actin content of HL-60 cells. Am J Physiol 261:C1154–C1161

Hammerman MR, Sacktor B, Daughaday WH (1980) myo-Inositol transport in renal brush border vesicles and it inhibition by D-glucose. Am J Physiol 239:F113–F120

Han X, Budreau AM, Chesney RW (1998) Molecular cloning and functional expression of an LLC-PK1 cell taurine transporter that is adaptively regulated by taurine. Adv Exp Med Biol 442:261–268

Handler JS, Kwon HM (2001) Cell and molecular biology of organic osmolyte accumulation in hypertonic renal cells. Nephron 87:106–110

Hardy SP, Goodfellow HR, Valverde MA, Gill DR, Sepulveda V, Higgins CF (1995) Protein kinase C-mediated phosphorylation of the human multidrug resistance P-glycoprotein regulates cell volume-activated chloride channels. EMBO J 14:68–75

Häussinger D (1996) The role of cellular hydration in the regulation of cell function. Biochem J 313:697–710

Häussinger D (1998) Osmoregulation of liver cell function: signalling, osmolytes and cell heterogeneity. Contrib Nephrol 123:185–204

Häussinger D, Hallbrucker C, Vom Dahl S, Decker S, Schweizer U, Lang F, Gerok W (1991) Cell volume is a major determinant of proteolysis control in liver. FEBS Lett 283:70–72

Hazama A, Okada Y (1988) Ca^{2+} sensitivity of volume-regulatory K^+ and Cl^- channels in cultured human epithelial cells. J Physiol 402:687–702

Hazama A, Okada Y (1990) Involvement of Ca^{2+}-induced Ca^{2+} release in the volume regulation of human epithelial cells exposed to a hypotonic medium. Biochem Biophys Res Commun 167:287–293

Hebert SC, Sun A (1988) Hypotonic cell volume regulation in mouse medullary thick ascending limb: effects of ADH. Am J Physiol 255:F962–F969

Heinzinger H, van den Boom F, Tinel H, Wehner F (2001) In rat hepatocytes, the hypertonic activation of Na^+ conductance and Na^+-K^+-$2Cl^-$ symport—but not Na^+-H^+ antiport—is mediated by protein kinase C. J Physiol (Lond) 536:703–715

Henson JH (1999) Relationships between the actin cytoskeleton and cell volume regulation. Microsc Res Tech 47:155–162

Hiki K, D'Andrea RJ, Furze J, Crawford J, Woollatt E, Sutherland GR, Vadas MA, Gamble JR (1999) Cloning, characterization, and chromosomal location of a novel human K^+-Cl^- cotransporter. J Biol Chem 274:10661–10667

Hirsch JR, Schlatter E (1997) Ca^{2+}-dependent K^+ channels in the cortical collecting duct of rat. Wien Klin Wochenschr 109:485–488

Hoffmann EK (1978) Regulation of cell volume by selective changes in the leak permeabilities of Ehrlich ascites tumor cells. In Jorgensen CB, Skadhauge E (eds) Osmotic and volume regulation. Alfred Benzon Symposium XI., Munksgaard, Copenhagen, pp. 397–417

Hoffmann EK, Dunham PB (1995) Membrane mechanisms and intracellular signalling in cell volume regulation. Int Rev Cytol 161:173–262

Hoffmann EK, Lambert IH (1994) On the similarity between the small Cl^- channel and the taurine channel activated after cell swelling in Ehrlich ascites tumor cells. Jpn J Physiol 44 Suppl 2:S49–S53

Hoffmann EK, Lambert IH, Simonsen LO (1988) Mechanisms in volume regulation in Ehrlich ascites tumor cells. Renal Physiol Biochem 11:221–247

Hoffmann EK, Mills JW (1999) Membrane events involved in volume regulation. Curr Top Membr 48:123–196

Hoffmann EK, Pedersen SF (1998) Sensors and signal transduction in the activation of cell volume regulatory ion transport systems. Contrib Nephrol 123:50–78

Hoffmann EK, Simonsen LO, Lambert IH (1984) Volume-induced increase of K^+ and Cl^- permeabilities in Ehrlich ascites tumor cells. Role of internal Ca^{2+}. J Membr Biol 78:211–222

Holtzman EJ, Kumar S, Faaland CA, Warner F, Logue PJ, Erickson SJ, Ricken G, Waldman J, Kumar S, Dunham PB (1998) Cloning, characterization, and gene organization of K-Cl cotransporter from pig and human kidney and C. elegans. Am J Physiol 275:F550–F564

Hoogerwerf WA, Tsao SC, Devuyst O, Levine SA, Yun CH, Yip JW, Cohen ME, Wilson PD, Lazenby AJ, Tse CM, Donowitz M (1996) NHE2 and NHE3 are human and rabbit intestinal brush-border proteins. Am J Physiol 270:G29–G41

Hooley R, Yu CY, Symons M, Barber DL (1996) G α 13 stimulates Na^+-H^+ exchange through distinct Cdc42-dependent and RhoA-dependent pathways. J Biol Chem 271:6152–6158

Horio M, Yamauchi A, Moriyama T, Imai E, Orita Y (1997) Osmotic regulation of amino acids and system A transport in Madin-Darby canine kidney cells. Am J Physiol 272:C804–C809

Huang R, Somjen GG (1997) Effects of hypertonia on voltage-gated ion currents in freshly isolated hippocampal neurons, and on synaptic currents in neurons in hippocampal slices. Brain Res 748:157–167

Huber SM, Gamper N, Lang F (2001) Chloride conductance and volume-regulatory nonselective cation conductance in human red blood cell ghosts. Pflugers Arch 441:551–558

Hubert MD, Levitan I, Hoffman MM, Zraggen M, Hofreiter ME, Garber SS (2000) Modulation of volume regulated anion current by I(Cln) Biochim Biophys Acta 1466:105–114

Hume JR, Duan D, Collier ML, Yamazaki J, Horowitz B (2000) Anion transport in heart. Physiol Rev 80:31–81

Hunter M (1990) Stretch-activated channels in the basolateral membrane of single proximal cells of frog kidney. Pflugers Arch 416:448–453

Hurst AM, Hunter M (1990) Stretch-activated channels in single early distal tubule cells of the frog. J Physiol (Lond) 430:13–24

Ingber DE (1997) Tensegrity: the architectural basis of cellular mechanotransduction. Annu Rev Physiol 59:575–599

Ishii TM, Silvia C, Hirschberg B, Bond CT, Adelman JP, Maylie J (1997) A human intermediate conductance calcium-activated potassium channel. Proc Natl Acad Sci USA 94:11651–11656

Isom LL, De Jongh KS, Catterall WA (1994) Auxiliary subunits of voltage-gated ion channels. Neuron 12:1183–1194

Itoh T, Yamauchi A, Miyai A, Yokoyama K, Kamada T, Ueda N, Fujiwara Y (1994) Mitogen-activated protein kinase and its activator are regulated by hypertonic stress in Madin-Darby canine kidney cells. J Clin Invest 93:2387–2392

Iwasa K, Tasaki I, Gibbons RC (1980) Swelling of nerve fibers associated with action potentials. Science 210:338–339

Jackson PS, Churchwell K, Ballatori N, Boyer JL, Strange K (1996) Swelling-activated anion conductance in skate hepatocytes: Regulation by cell Cl^- and ATP. Am J Physiol 270:C57–C65

Jackson PS, Morrison R, Strange K (1994) The volume-sensitive organic osmolyte-anion channel VSOAC is regulated by nonhydrolytic ATP binding. Am J Physiol 267:C1203–C1209

Jackson PS, Strange K (1993) Volume-sensitive anion channels mediate swelling-activated inositol and taurine efflux. Am J Physiol 265:C1489–C1500

Jackson PS, Strange K (1995a) Characterization of the voltage-dependent properties of a volume-sensitive anion conductance. J Gen Physiol 105:661–677

Jackson PS, Strange K (1995b) Single-channel properties of a volume-sensitive anion conductance. Current activation occurs by abrupt switching of closed channels to an open state. J Gen Physiol 105:643-660

Jacobsen JG, Smith LH (1968) Biochemistry and physiology of taurine and taurine derivatives. Physiol Rev 48:424–511

Jalonen T (1993) Single-channel characteristics of the large-conductance anion channel in rat cortical astrocytes in primary culture. Glia 9:227–237

Jarolimek W, Lewen A, Misgeld U (1999) A furosemide-sensitive K^+-Cl^- cotransporter counteracts intracellular Cl^- accumulation and depletion in cultured rat midbrain neurons. J Neurosci 19:4695–4704

Jensen BS, Jessen F, Hoffmann EK (1993) Na^+, K^+, Cl^- cotransport and its regulation in Ehrlich ascites tumor cells. Ca^{2+}/calmodulin and protein kinase C dependent pathways. J Membr Biol 131:161–178

Jensen BS, Strobaek D, Christophersen P, Jorgensen TD, Hansen C, Silahtaroglu A, Olesen SP, Ahring PK (1998) Characterization of the cloned human intermediate-conductance Ca^{2+}-activated K^+ channel. Am J Physiol 275:C848–C856

Jensen BS, Strobaek D, Olesen SP, Christophersen P (2001) The Ca^{2+}-activated K^+ channel of intermediate conductance: a molecular target for novel treatments? Curr Drug Targets 2:401–422

Jentsch TJ (1996) Chloride channels: A molecular perspective. Curr Opin Neurobiol 6:303–310

Jentsch TJ, Friedrich T, Schriever A, Yamada H (1999) The CLC chloride channel family. Pflugers Arch 437:783–795

Jentsch TJ, Steinmeyer K, Schwarz G (1990) Primary structure of Torpedo marmorata chloride channel isolated by expression cloning in Xenopus oocytes. Nature 348:510–514

Jiang L, Chernova MN, Alper SL (1997) Secondary regulatory volume increase conferred on Xenopus oocytes by expression of AE2 anion exchanger. Am J Physiol 272, C191-C202

Joiner WJ, Wang LY, Tang MD, Kaczmarek LK (1997) hSK4 a member of a novel subfamily of calcium-activated potassium channels. Proc Natl Acad Sci USA 94:11013–11018

Jorgensen NK, Christensen S, Harbak H, Brown AM, Lambert IH, Hoffmann EK, Simonsen LO (1997) On the role of calcium in the regulatory volume decrease (RVD) response in Ehrlich mouse ascites tumor cells. J Membr Biol 157:281–299

Jorgensen NK, Lambert IH, Hoffmann EK (1996) Role of LTD4 in the regulatory volume decrease response in Ehrlich ascites tumor cells. J Membr Biol 151:159–173

Jorgensen NK, Petersen SF, Hoffmann EK (1999) Thrombin-, bradykinin-, and arachidonic acid-induced Ca^{2+} signaling in Ehrlich ascites tumor cells. Am J Physiol 276:C26–C37

Junankar PR, Kirk K (2000) Organic osmolyte channels: a comparative view. Cell Physiol Biochem 10:355–360

Kajimura M, O'Donnell ME, Curry FE (1997) Effect of cell shrinkage on permeability of cultured bovine aortic endothelia and frog mesenteric capillaries. J Physiol 503:413–425

Kanli H, Norderhus E (1998) Cell volume regulation in proximal renal tubules from trout (Salmo trutta) J Exp Biol 201 (Pt 9):1405–1419

Kashani AH, Chen BM, Grinnell AD (2001) Hypertonic enhancement of transmitter release from frog motor nerve terminals: Ca^{2+} independence and role of integrins. J Physiol 530:243–252

Kawahara K (1990) A stretch-activated K^+ channel in the basolateral membrane of Xenopus kidney proximal tubule cells. Pflugers Arch 415:624–629

Kawahara K, Ogawa A, Suzuki M (1991) Hypo-osmotic activation of Ca-activated K channels in cultured rabbit kidney proximal tubule cells. Am J Physiol 260:F27–F33

Kempson SA (1998) Differential activation of system A and betaine/GABA transport in MDCK cell membranes by hypertonic stress. Biochim Biophys Acta 1372:117–123

Khanna R, Chang MC, Joiner WJ, Kaczmarek LK, Schlichter LC (1999) hSK4/hIK1 a calmodulin-binding KCa channel in human T lymphocytes. Roles in proliferation and volume regulation. J Biol Chem 274:14838–14849

Kilic G, Fitz JG (2002) Heterotrimeric G-proteins activate Cl- channels through stimulation of a cyclooxygenase-dependent pathway in a model liver cell line. J Biol Chem 277:11721–11727

Kim D (1993) Novel cation-selective mechanosensitive ion channel in the atrial cell membrane. Circ Res 72:225–231

Kim D, Fu C (1993) Activation of a nonselective cation channel by swelling in atrial cells. J Membr Biol 135:27–37

Kim D, Sladek CD, Aguado-Velasco C, Mathiasen JR (1995) Arachidonic acid activation of a new family of K^+ channels in cultured rat neuronal cells. J Physiol 484 (Pt 3):643–660

Kim RD, Roth TP, Darling CE, Ricciardi R, Schaffer BK, Chari RS (2001) Hypo-osmotic stress stimulates growth in HepG2 cells via protein kinase B-dependent activation of activator protein-1. J Gastrointest Surg 5:546–555

Kimura K, Ito M, Amano M, Chihara K, Fukata Y, Nakafuku M, Yamamori B, Feng J, Nakano T, Okawa K, Iwamatsu A, Kaibuchi K (1996) Regulation of myosin phosphatase by Rho and Rho-associated kinase (Rho-kinase) Science 273:245–248

Kinne RKH (1993) The role of organic osmolytes in osmoregulation: from bacteria to mammals. J Exp Zool 265:346–355

Kinne RKH (1998) Mechanisms of osmolyte release. Contrib Nephrol 123:34–49

Kinne RKH, Czekay RP, Grunewald JM, Mooren FC, Kinne-Saffran E (1993) Hypotonicity-evoked release of organic osmolytes from distal renal cells: systems, signals, and sidedness. Renal Physiol Biochem 16:66–78

Kinne RKH, Boese SH, Kinne-Saffran E, Ruhfus B, Tinel H, Wehner F (1996) Osmoregulation in the renal papilla: membranes, messengers and molecules. Kidney Int 49:1686–1689

Kinne RKH, Grunewald RW, Ruhfus B, Kinne-Saffran E (1997) Biochemistry and physiology of carbohydrates in the renal collecting duct. J Exp Zool 279:436–442

Kinne RKH, Tinel H, Kipp H, Kinne-Saffran E (2000) Regulation of sorbitol efflux in different renal medullary cells: similarities and diversities. Cell Physiol Biochem 10:371–378

Kinne RKH, Kipp H, Ruhfus B, Wehner F, Boese SH, Kinne-Saffran E (2001) Organic osmolyte channels in the renal medulla: Their properties and regulation. Am Zool 41:728–733

Kinne RKH, Ruhfus B, Tinel H, Boese SH, Wehner F, Kinne–Saffran E (1995) Renal organic osmolytes: signal transduction pathways and release mechanisms. In de Santo NG, Capasso G (eds.) Acid-base and electrolyte balance. Molecular, cellular and clinical aspects. Istituto Italiano per gli Studi Filosofici, Cosenza, pp. 237–242

Kinne-Saffran E, Kinne RKH (1997) Sorbitol uptake in plasma membrane vesicles isolated from immortalized rabbit TALH cells: activation by a Ca^{2+}/calmodulin-dependent protein kinase. J Membr Biol 159:231–238

Kinnunen PK (2000) Lipid bilayers as osmotic response elements. Cell Physiol Biochem 10:243–250

Kirber MT, Ordway RW, Clapp LH, Walsh Jr. JV, Singer JJ (1992) Both membrane stretch and fatty acids directly activate large conductance Ca^{2+}-activated K^+ channels in vascular smooth muscle cells. FEBS Lett 297:24–28

Kirk K (1997) Swelling-activated organic osmolyte channels. J. Membr. Biol. 158:1–16

Kirk K, Strange K (1998) Functional properties and physiological roles of organic solute channels. Annu Rev Physiol 60:719–739

Kirschner U, Tinel H, Rosin-Steiner S, Giffey A, Kinne RKH, Wehner F (1998) Single rat hepatocytes in primary culture as a model system for the study of regulatory volume increase (RVI) in liver. Nova Acta Leopold 306:299–303

Kirschner U, Van Driessche W, Werner A, Wehner F (2003) Hypertonic activation of phospholemman in solitary rat hepatocytes in primary culture. FEBS Lett 537:151-156

Kizer N, Harter L, Hruska K, Alvarez U, Duncan R (1999) Volume regulatory decrease in UMR-106.01 cells is mediated by specific $\alpha 1$ subunits of L-type calcium channels. Cell Biochem Biophys 31:65–79

Klanke CA, Su YR, Callen DF, Wang Z, Meneton P, Baird N, Kandasamy RA, Orlowski J, Otterud BE, Leppert M, (1995) Molecular cloning and physical and genetic mapping of a novel human Na^+/H^+ exchanger (NHE5/SLC9A5) to chromosome 16q22.1. Genomics 25:615–622

Klein JD, Lamitina ST, O'Neill WC (1999) JNK is a volume-sensitive kinase that phosphorylates the Na-K-2Cl cotransporter in vitro. Am J Physiol 277:C425–C431

Klein M, Seeger P, Schuricht B, Alper SL, Schwab A (2000) Polarization of Na^+/H^+ and Cl^-/HCO_3^- exchangers in migrating renal epithelial cells. J Gen Physiol 115:599–608

Kobayashi T, Cohen P (1999) Activation of serum- and glucocorticoid-regulated protein kinase by agonists that activate phosphatidylinositide 3-kinase is mediated by 3-phosphoinositide-dependent protein kinase-1 (PDK1) and PDK2. Biochem J 339:319–328

Koch JP, Korbmacher C (1999) Osmotic shrinkage activates nonselective cation (NSC) channels in various cell types. J Membr Biol 168:131–139

Koch JP, Korbmacher C (2000) Mechanism of shrinkage activation of nonselective cation channels in M-1 mouse cortical collecting duct cells. J Membr Biol 177:231–242

Köhler M, Hirschberg B, Bond CT, Kinzie JM, Marrion NV, Maylie J, Adelman JP (1996) Small-conductance, calcium-activated potassium channels from mammalian brain. Science 273:1709–1714

Koomoa DL, Musch MW, Goldstein L (2002) Comparison of the osmolyte transport properties induced by trAE1 versus IClswell in *Xenopus* oocytes. J Membr Biol 185:57–63

Kopito RR, Lodish HF (1985) Primary structure and transmembrane orientation of the murine anion exchange protein. Nature 316:234–238

Korbmacher C, Volk T, Segal AS, Boulpaep EL, Frömter E (1995) A calcium-activated and nucleotide-sensitive nonselective cation channel in M-1 mouse cortical collecting duct cells. J Membr Biol 146:29–45

Kowdley GC, Ackerman SJ, Chen Z, Szabo G, Jones LR, Moorman JR (1997) Anion, cation, and zwitterion selectivity of phospholemman channel molecules. Biophys J 72:141–145

Krapivinsky GB, Ackerman MJ, Gordon EA, Krapivinsky LD, Clapham DE (1994) Molecular characterization of a swelling-induced chloride conductance regulatory protein, pI_{Cln}. Cell 76:439–448

Krause U, Rider MH, Hue L (1996) Protein kinase signaling pathway triggered by cell swelling and involved in the activation of glycogen synthase and acetyl-CoA carboxylase in isolated rat hepatocytes. J Biol Chem 271:16668–16673

Kregenow FM, Robbie DE, Orloff J (1976) Effect of norepinephrine and hypertonicity on K influx and cyclic AMP in duck erythrocytes. Am J Physiol 231:306–311

Kubitz R, Warth R, Allert N, Kunzelmann K, Greger R (1992) Small-conductance chloride channels induced by cAMP, Ca^{2+}, and hypotonicity in HT29 cells: ion selectivity, additivity and stilbene sensitivity. Pflugers Arch 421:447–454

Kulanthaivel P, Cool DR, Ramamoorthy S, Mahesh VB, Leibach FH, Ganapathy V (1991) Transport of taurine and its regulation by protein kinase C in the JAR human placental choriocarcinoma cell line. Biochem J 277 (Pt 1), 53–58

Kurashima K, D'Souza S, Szászi K, Ramjeesingh R, Orlowski J, Grinstein S (1999) The apical Na^+/H^+ exchanger isoform NHE3 is regulated by the actin cytoskeleton. J Biol Chem 274:29843–29849

Kwon HM, Yamauchi A, Uchida S, Preston AS, Garcia-Perez A, Burg MB, Handler JS (1992) Cloning of the cDNa for a Na^+/myo-inositol cotransporter a hypertonicity stress protein. J Biol Chem 267:6297–6301

Laich A, Gschwentner M, Krick W, Nagl UO, Fürst J, Hofer S, Susanna A, Schmarda A, Deetjen P, Burckhardt G, Paulmichl M (1997) I_{Cln} a chloride channel cloned from kidney cells, is activated during regulatory volume decrease. Kidney Int 51:477–478

Lambert IH (1989) Leukotriene-D4 induced cell shrinkage in Ehrlich ascites tumor cells. J Membr Biol 108:165–176

Lambert IH, Hoffmann EK (1993) Regulation of taurine transport in Ehrlich ascites tumor cells. J Membr Biol 131:67–79

Lambert IH, Hoffmann EK (1994) Cell swelling activates separate taurine and chloride channels in Ehrlich mouse ascites tumor cells. J Membr Biol 142:289–298

Lambert IH, Hoffmann EK, Christensen P (1987) Role of prostaglandins and leukotrienes in volume regulation by Ehrlich ascites tumor cells. J Membr Biol 98:247–256

Lang F, Busch GL, Ritter M, Völkl H, Waldegger S, Gulbins E, Häussinger D (1998a) Functional significance of cell volume regulatory mechanisms. Physiol Rev 78:247–306

Lang F, Busch GL, Völkl H (1998b) The diversity of volume regulatory mechanisms. Cell Physiol Biochem 8:1–45

Lang F, Stehle T, Häussinger D (1989) Water, K^+, H^+, lactate and glucose fluxes during cell volume regulation in perfused rat liver. Pflugers Arch 413:209–216

Lansman JB, Hallam TJ, Rink TJ (1987) Single stretch-activated ion channels in vascular endothelial cells as mechanotransducers? Nature 325:811–813

Lapointe J-Y, Garneau L, Bell PD, Cardinal J (1990) Membrane crosstalk in the mammalian proximal tubule during alterations in transepithelial sodium transport. Am J Physiol 258:F339–F345

Larsen AK, Jensen BS, Hoffmann EK (1994) Activation of protein kinase C during cell volume regulation in Ehrlich mouse ascites tumor cells. Biochim Biophys Acta 1222:477–482

Lauf PK, Adragna NC (2000) K-Cl cotransport: properties and molecular mechanism. Cell Physiol Biochem 10:341–354

Lauf PK, Zhang J, Gagnon KB, Delpire E, Fyffe RE, Adragna NC (2001) K-Cl cotransport: immunohistochemical and ion flux studies in human embryonic kidney (HEK293) cells transfected with full-length and C- terminal-domain-truncated KCC1 cDNAs. Cell Physiol Biochem 11:143–160

Law RO (1998) The role of taurine in the regulation of brain cell volume in chronically hyponatraemic rats. Neurochem Int 33:467–472

Law RO (1999) Amino acid efflux and cell volume regulation in cerebrocortical minislices prepared from chronically hyponatraemic and hypernatraemic rats. Neurochem Int 35:423–430

Lawonn P, Hoffmann EK, Hougaard C, Wehner F (2003) A cell shrinkage-induced cation conductance with a novel pharmacology in Ehrlich-Lettre-ascites tumour cells. FEBS Lett (in press)

Leibowich S, DeLong J, Civan MM (1988) Apical Na^+ permeability of frog skin during serosal Cl^- replacement. J Membr Biol 102:121–130

Lemonnier L, Prevarskaya N, Shuba Y, Vanden Abeele F, Nilius B, Mazurier J, Skryma R (2002) Ca^{2+} modulation of volume-regulated anion channels: evidence for colocalization with store-operated channels. FASEB J 16:222–224

Lesage F, Lazdunski M (2000) Molecular and functional properties of two-pore-domain potassium channels. Am J Physiol 279:F793–F801

Levitan IB (1994) Modulation of ion channels by protein phosphorylation and dephosphorylation. Annu Rev Physiol 56:193–212

Lewis RA, Bursell JD, Kirk K (1996) Anion-selectivity of the swelling-activated osmolyte channel in eel erythrocytes. J Membr Biol 149:103–111

Lewis RS, Cahalan MD (1995) Potassium and calcium channels in lymphocytes. Annu Rev Immunol 13:623–653

Lewis RS, Ross PE, Cahalan MD (1993) Chloride channels activated by osmotic stress in T lymphocytes. J Gen Physiol 101:801–826

Lewis SA, Butt AG, Bowler MJ, Leader JP, Macknight ADC (1985) Effects of anions on cellular volume and transepithelial Na^+ transport across toad urinary bladder. J Membr Biol 83:119–137

Li CH, Breton S, Morrison R, Cannon CL, Emma F, Sanchez-Olea R, Bear C, Strange K (1998) Recombinant pI_{Cln} forms highly cation-selective channels when reconstituted into artificial and biological membranes. J Gen Physiol 112:727–736

Li G, Liu Y, Olson JE (2002) Calcium/calmodulin-modulated chloride and taurine conductances in cultured rat astrocytes. Brain Res 925:1–8

Li Q, Jungmann V, Kiyatkin A, Low PS (1996) Prostaglandin E2 stimulates a Ca^{2+}-dependent K^+ channel in human erythrocytes and alters cell volume and filterability. J Biol Chem 271:18651–18656

Liedtke CM, Cole TS (2002) Activation of NKCC1 by hyperosmotic stress in human tracheal epithelial cells involves PKC-delta and ERK. Biochim Biophys Acta 1589:77–88

Liedtke W, Choe Y, Marti-Renom MA, Bell AM, Denis CS, Sali A, Hudspeth AJ, Friedman JM, Heller S (2000) Vanilloid receptor-related osmotically activated channel (VR-OAC) a candidate vertebrate osmoreceptor. Cell 103:525–535

Light DB, Mertins TM, Belongia JA, Witt CA (1997) 5-Lipoxygenase metabolites of arachidonic acid regulate volume decrease by mudpuppy red blood cells. J Membr Biol 158:229–239

Lima L, Cubillos S, Guerra A (2000) Regulation of high affinity taurine transport in goldfish and rat retinal cells. Adv Exp Med Biol 483:431–440

Ling BN, Webster CL, Eaton DC (1992) Eicosanoids modulate apical Ca^{2+}-dependent K^+ channels in cultured rabbit principal cells. Am J Physiol 263:F116–F126

Lock H, Valverde MA (2000) Contribution of the IsK (MinK) potassium channel subunit to regulatory volume decrease in murine tracheal epithelial cells. J Biol Chem 275:34849–34852

Logsdon NJ, Kang J, Togo JA, Christian EP, Aiyar J (1997) A novel gene, hKCa4, encodes the calcium-activated potassium channel in human T lymphocytes. J Biol Chem 272:32723–32726

Low SY, Rennie MJ, Taylor PM (1996) Modulation of glycogen synthesis in rat skeletal muscle by changes in cell volume. J Physiol 495:299–303

Low SY, Taylor PM (1998) Integrin and cytoskeletal involvement in signalling cell volume changes to glutamine transport in rat skeletal muscle. J Physiol 512:481–485

Luby-Phelps K (2000) Cytoarchitecture and physical properties of cytoplasm: volume, viscosity, diffusion, intracellular surface area. Int Rev Cytol 192:189–221

Lucke B, McCutcheon M (1932) The living cell as an osmotic system and its permeability to water. Physiol Rev 12:68–139

Lundgren DW, Moore RM, Collins PL, Moore JJ (1997) Hypotonic stress increases cyclooxygenase-2 expression and prostaglandin release from amnion-derived WISH cells. J Biol Chem 272:20118–20124

Ma TH, Verkman AS (1999) Aquaporin water channels in gastrointestinal physiology. J Physiol (Lond) 517:317–326

MacLeod RJ, Hamilton JR (1999a) Ca^{2+}/Calmodulin kinase II and decreases in intracellular pH are required to activate K^+ channels after substantial swelling in villus epithelial cells. J Membr Biol 172:59–66

MacLeod RJ, Hamilton JR (1999b) Increases in intracellular pH and Ca^{2+} are essential for K^+ channel activation after modest "physiological" swelling in villus epithelial cells. J Membr Biol 172:47–58

Macri P, Breton S, Marsolais M, Lapointe JY, Laprade R (1997) Hypertonicity decreases basolateral K^+ and Cl^- conductances in rabbit proximal convoluted tubule. J Membr Biol 155:229–237

Mähler S, Kinne-Saffran E, Fujisue H, Kinne RKH, Föllmann W (1998) Regulation of sorbitol content in cultured porcine urinary bladder epithelial cells. Am J Physiol 274:F342–F347

Marchenko SM, Sage SO (2000) Hyperosmotic but not hypo-osmotic stress evokes a rise in cytosolic Ca^{2+} concentration in endothelium of intact rat aorta. Exp Physiol 85:151–157

Margalit A, Livne AA, Funder J, Granot Y (1993a) Initiation of RVD response in human platelets: mechanical-biochemical transduction involves pertussis-toxin-sensitive G protein and phospholipase A2. J Membr Biol 136:303–311

Margalit A, Sofer Y, Grossman S, Reynaud D, Pace-Asciak CR, Livne AA (1993b) Hepoxilin A3 is the endogenous lipid mediator opposing hypotonic swelling of intact human platelets. Proc Natl Acad Sci USA 90:2589–2592

Marinelli RA, LaRusso NF (1997) Aquaporin water channels in liver: Their significance in bile formation. Hepatology 26:1081–1084

Marunaka Y, Niisato N, O'Brodovich H, Post M, Tanswell AK (1999) Roles of Ca^{2+} and protein tyrosine kinase in insulin action on cell volume via Na^+ and K^+ channels and $Na^+/K^+/2Cl^-$ cotransporter in fetal rat alveolar type II pneumocyte. J Membr Biol 168 91–101

Mastrocola T, Lambert IH, Kramhoft B, Rugolo M, Hoffmann EK (1993) Volume regulation in human fibroblasts: role of Ca^{2+} and 5-lipoxygenase products in the activation of the Cl^- efflux. J Membr Biol 136:55–62

Matskevitch I, Wagner CA, Stegen C, Broer S, Noll B, Risler T, Kwon HM, Handler JS, Waldegger S, Busch AE, Lang F (1999) Functional characterization of the Betaine/gamma-aminobutyric acid transporter BGT-1 expressed in *Xenopus* oocytes. J Biol Chem 274:16709–16716

Matsuoka Y, Yamauchi A, Nakanishi T, Sugiura T, Kitamura H, Horio M, Takamitsu Y, Ando A, Imai E, Hori M (1999) Response to hypertonicity in mesothelial cells: role of Na^+/myo-inositol cotransporter. Nephrol Dial Transplant 14:1217–1223

Maunsbach AB, Marples D, Chin E, Ning G, Bondy C, Agre P, Nielsen S (1997) Aquaporin-1 water channel expression in human kidney. J Am Soc Nephrol 8:1–14

McCarty NA, O'Neil RG (1991) Calcium-dependent control of volume regulation in renal proximal tubule cells: I. Swelling-activated Ca^{2+} entry and release. J Membr Biol 123:149–160

Meier R, Thelen M, Hemmings BA (1998) Inactivation and dephosphorylation of protein kinase Balpha (PKBalpha) promoted by hyperosmotic stress. EMBO J 17:7294–7303

Mercado A, Song LY, Vázquez N, Mount DB, Gamba G (2000) Functional comparison of the K^+-Cl^- cotransporters KCC1 and KCC4. J Biol Chem 275:30326–30334

Meyer K, Korbmacher C (1996) Cell swelling activates ATP-dependent voltage-gated chloride channels in M-1 mouse cortical collecting duct cells. J Gen Physiol 108:177–193

Mignen O, Le Gall C, Harvey BJ, Thomas S (1999) Volume regulation following hypotonic shock in isolated crypts of mouse distal colon. J Physiol 515:501–510

Minton AP (1994) Influence of macromolecular crowding on intracellular association reactions: possible role in volume regulation. In Strange K (ed) Cellular and molecular physiology of cell volume regulation. CRC Press, Boca Raton, pp 181–190

Minton AP, Colclasure GC, Parker JC (1992) Model for the role of macromolecular crowding in regulation of cellular volume. Proc Natl Acad Sci USA 89:10504–10506

Mitchell CH, Zhang JJ, Wang L, Jacob TJ (1997) Volume-sensitive chloride current in pigmented ciliary epithelial cells: role of phospholipases. Am J Physiol 272:C212–C222
Miyai A, Yamauchi A, Moriyama T, Kaneko T, Takenaka M, Sugiura T, Kitamura H, Ando A, Tohyama M, Shimada S, Imai E, Kamada T (1996) Expression of betaine transporter mRNA: its unique localization and rapid regulation in rat kidney. Kidney Int 50:819–827
Miyauchi A, Notoya K, Mikuni-Takagaki Y, Takagi Y, Goto M, Miki Y, Takano-Yamamoto T, Jinnai K, Takahashi K, Kumegawa M, Chihara K, Fujita T (2000) Parathyroid hormone-activated volume-sensitive calcium influx pathways in mechanically loaded osteocytes. J Biol Chem 275:3335–3342
Moeckel GW, Lai LW, Guder WG, Kwon HM, Lien YH (1997) Kinetics and osmoregulation of Na^+- and Cl^--dependent betaine transporter in rat renal medulla. Am J Physiol 272:F100–F106
Mollerup J, Lambert IH (1998) Calyculin A modulates the kinetic constants for the Na^+-coupled taurine transport in Ehrlich ascites tumor cells. Biochim Biophys Acta 1371:335–344
Mongin AA, Cai Z, Kimelberg HK (1999) Volume-dependent taurine release from cultured astrocytes requires permissive $[Ca^{2+}]_i$ and calmodulin. Am J Physiol 277:C823–C832
Montell C, Rubin GM (1989) Molecular characterization of the *Drosophila* trp locus: a putative integral membrane protein required for phototransduction. Neuron 2:1313–1323
Montrose-Rafizadeh C, Guggino WB (1991) Role of intracellular calcium in volume regulation by rabbit medullary thick ascending limb cells. Am J Physiol 260:F402–F409
Montrose-Rafizadeh C, Guggino WB, Montrose MH (1991) Cellular differentiation regulates expression of Cl^- transport and cystic fibrosis transmembrane conductance regulator mRNA in human intestinal cells. J Biol Chem 266:4495–4499
Mooren FC, Kinne RKH (1994) Intracellular calcium in primary cultures of rat renal inner medullary collecting duct cells during variations of extracellular osmolality. Pflugers Arch 427:463–472
Moorman JR, Ackerman SJ, Kowdley GC, Griffin MP, Mounsey JP, Chen ZH, Cala SE, O'Brian JJ, Szabo G, Jones LR (1995) Unitary anion currents through phospholemman channel molecules. Nature 377:737–740
Moorman JR, Jones LR (1998) Phospholemman: A cardiac taurine channel involved in regulation of cell volume. Adv Exp Med Biol 442:219–228
Moorman JR, Palmer CJ, John III JE, Durieux ME Jones LR (1992) Phospholemman expression induces a hyperpolarization-activated chloride current in Xenopus oocytes. J Biol Chem 267:14551–14554
Morales-Mulia M, Pasantes-Morales H, Morán J (2000) Volume sensitive efflux of taurine in HEK293 cells overexpressing phospholemman. Biochim Biophys Acta 1496:252–260
Morales-Mulia S, Cardin V, Torres-Marquez ME Crevenna A, Pasantes-Morales H (2001) Influence of protein kinases on the osmosensitive release of taurine from cerebellar granule neurons. Neurochem Int 38:153–161
Morán J, Morales-Mulia M, Pasantes-Morales H (2001) Reduction of phospholemman expression decreases osmosensitive taurine efflux in astrocytes. Biochim Biophys Acta 1538:313–320
Morris CE (1990) Mechanosensitive ion channels. J Membr Biol 113:93–107
Motais R, Guizouarn H, Garcia-Romeu F (1991) Red cell volume regulation: the pivotal role of ionic strength in controlling swelling-dependent transport systems. Biochim Biophys Acta 1075:169–180
Motais R, Fievet B, Borgese F, Garcia-Romeu F (1997) Association of the band 3 protein with a volume-activated, anion and amino acid channel: a molecular approach. J Exp Biol 200:361–367
Mounsey JP, Lu KP, Patel MK, Chen ZH, Horne LT, John JE, III, Means AR, Jones LR, Moorman JR (1999) Modulation of Xenopus oocyte-expressed phospholemman-induced ion currents by co-expression of protein kinases. Biochim Biophys Acta 1451:305–318
Mount DB, Mercado A, Song L, Xu J, George AL, Jr., Delpire E, Gamba G (1999) Cloning and characterization of KCC3 and KCC4, new members of the cation-chloride cotransporter gene family. J Biol Chem 274:16355–16362
Mountian I, Chou KY, van Driessche W (1996) Electrolyte transport mechanisms involved in regulatory volume increase in C6 glioma cells. Am J Physiol 271:C1041–C1048
Musch MW, Goldstein L (1990) Hypotonicity stimulates phosphatidylcholine hydrolysis and generates diacylglycerol in erythrocytes. J Biol Chem 265:13055–13059
Musch MW, Leffingwell TR, Goldstein L (1994) Band 3 modulation and hypotonic-stimulated taurine efflux in skate erythrocytes. Am J Physiol 266:R65–R74
Musch MW, Luer CA, Davis-Amaral EM, Goldstein L (1997) Hypotonic stress induces translocation of the osmolyte channel protein pICln in embryonic skate (Raja eglanteria) heart. J Exp Zool 277:460–463
Musch MW, Davis-Amaral EM, Vandenburgh HH, Goldstein L (1998) Hypotonicity stimulates translocation of ICln in neonatal rat cardiac myocytes. Pflugers Arch 436:415–422
Muto S, Ohtaka A, Nemoto J, Kawakami K, Asano Y (1998) Effects of hyperosmolality on Na, K-ATPase gene expression in vascular smooth muscle cells. J Membr Biol 162:233–245

Nagasaki M, Ye L, Duan D, Horowitz B, Hume JR (2000) Intracellular cyclic AMP inhibits native and recombinant volume-regulated chloride channels from mammalian heart. J Physiol 523:705–717

Nakamura H, Huang SH, Takakura K (1996) High-affinity taurine uptake and its regulation by protein kinase C in human glioma cells. Adv Exp Med Biol 403:377–384

Nakanishi T, Balaban RS, Burg MB (1988) Survey of osmolytes in renal cell lines. Am J Physiol 255:C181–C191

Nakanishi T, Turner RJ, Burg MB (1989) Osmoregulatory changes in myo-inositol transport by renal cells. Proc Natl Acad Sci USA 86:6002–6006

Nakanishi T, Turner RJ, Burg MB (1990) Osmoregulation of betaine transport in mammalian renal medullary cells. Am J Physiol 258:F1061–F1067

Napathorn S, Spring KR (1994) Further characterization of the sorbitol permease in PAP-HT25 cells. Am J Physiol 267:C514–C519

Naray-Fejes-Toth A, Canessa C, Cleaveland ES, Aldrich G, Fejes-Toth G (1999) sgk is an aldosterone-induced kinase in the renal collecting duct. Effects on epithelial Na^+ channels. J Biol Chem 274:16973–16978

Nath SK, Hang CY, Levine SA, Yun CH, C., Montrose MH, Donowitz M, Tse CM (1996) Hyperosmolarity inhibits the Na^+/H^+ exchanger isoforms NHE2 and NHE3: An effect opposite to that on NHE1. Am J Physiol 270:G431–G441

Nelson DJ, Tien XY, Xie WW, Brasitus TA, Kaetzel MA, Dedman JR (1996) Shrinkage activates a nonselective conductance: involvement of a Walker-motif protein and PKC. Am J Physiol 270:C179–C191

Niemeyer MI, Cid LP, Barros LF, Sepulveda FV (2001) Modulation of the two-pore domain acid-sensitive K^+ channel TASK-2 (KCNK5) by changes in cell volume. J Biol Chem 276:43166–43174

Niggli V, Andreoli C, Roy C, Mangeat P (1995) Identification of a phosphatidylinositol-4,5-bis-phosphate-binding domain in the N-terminal region of ezrin. FEBS Lett 376:172–176

Nilius B, Droogmans G (2001) Ion channels and their functional role in vascular endothelium. Physiol Rev 81:1415–1459

Nilius B, Sehrer J, Droogmans G (1994a) Permeation properties and modulation of volume-activated Cl^--currents in human endothelial cells. Br J Pharmacol 112:1049–1056

Nilius B, Sehrer J, Viana F, De Greef C, Raeymaekers L, Eggermont J, Droogmans G (1994b) Volume-activated Cl^- currents in different mammalian nonexcitable cell types. Pflugers Arch 428:364–371

Nilius B, Sehrer J, De Smet P, van Driessche W, Droogmans G (1995) Volume regulation in a toad epithelial cell line: Role of coactivation of K^+ and Cl^- channels. J Physiol (Lond) 487:357–378

Nilius B, Eggermont J, Voets T, Buyse G, Manolopoulos V, Droogmans G (1997) Properties of volume-regulated anion channels in mammalian cells. Prog Biophys Mol Biol 68:69–119

Nilius B, Prenen J, Voets T, Eggermont J, Droogmans G (1998) Activation of volume-regulated chloride currents by reduction of intracellular ionic strength in bovine endothelial cells. J Physiol 506:353–361

Nilius B, Voets T, Prenen J, Barth H, Aktories K, Kaibuchi K, Droogmans G, Eggermont J (1999) Role of Rho and Rho kinase in the activation of volume-regulated anion channels in bovine endothelial cells. J Physiol (Lond) 516:67–74

Nilius B, Prenen J, Wissenbach U, Bodding M, Droogmans G (2001) Differential activation of the volume-sensitive cation channel TRP12 (OTRPC4) and volume-regulated anion currents in HEK-293 cells. Pflugers Arch 443:227–233

Noel J, Roux D, Pouysségur J (1996) Differential localization of Na^+/H^+ exchanger isoforms (NHE1 and NHE3) in polarized epithelial cell lines. J Cell Sci 109:929–939

Noulin JF, Brochiero E, Lapointe JY, Laprade R (1999) Two types of K^+ channels at the basolateral membrane of proximal tubule: inhibitory effect of taurine. Am J Physiol 277:F290–F297

Novak JE, Agranoff BW, Fisher SK (2000) Regulation of Myo-inositol homeostasis in differentiated human NT2-N neurons. Neurochem Res 25:561–566

O'Donnell ME, Martinez A, Sun D (1995) Endothelial Na-K-Cl cotransport regulation by tonicity and hormones: phosphorylation of cotransport protein. Am J Physiol 269:C1513–C1523

O'Flaherty L, Stapleton PP, Redmond HP, Bouchier-Hayes DJ (1997) Intestinal taurine transport: a review. Eur J Clin Invest 27:873–880

O'Neil RG, Leng L (1997) Osmo-mechanically sensitive phosphatidylinositol signaling regulates a Ca^{2+} influx channel in renal epithelial cells. Am J Physiol 273:F120–F128

Ohta M, Tanimoto T, Tanaka A (1990) Characterization of aldose reductase and aldehyde reductase from the medulla of rat kidney. Chem Pharm Bull (Tokyo) 38:1639–1643

Ohtaka A, Muto S, Nemoto J, Kawakami K, Nagano K, Asano Y (1996) Hyperosmolality stimulates Na-K-ATPase gene expression in inner medullary collecting duct cells. Am J Physiol 270:F728–F738

Ohtsuyama M, Suzuki Y, Samman G, Sato F, Sato K (1993) Cell volume analysis of gramicidin-treated eccrine clear cells to study regulation of Cl channels. Am J Physiol 265:C1090–C1099

Oike M, Droogmans G, Nilius B (1994) The volume-activated chloride current in human endothelial cells depends on intracellular ATP. Pflugers Arch 427:184–186

Oike M, Kimura C, Koyama T, Yoshikawa M, Ito Y (2000) Hypotonic stress-induced dual Ca^{2+} responses in bovine aortic endothelial cells. Am J Physiol Heart Circ Physiol 279:H630–H638

Oiki S, Kubo M, Okada Y (1994) Mg^{2+} and ATP-dependence of volume-sensitive Cl^- channels in human epithelial cells. Jpn J Physiol 44: S77–S79

Okada Y (1997) Volume expansion-sensing outward-rectifier Cl^- channel: fresh start to the molecular identity and volume sensor. Am J Physiol 273:C755–C789

Okada Y (1998) Cell volume-sensitive chloride channels. Contrib Nephrol 123:21–33

Okada Y, Hazama A (1989) Volume-regulatory ion channels in epithelial cells. News Physiol Sci 4:238–242

Okada Y, Oiki S, Hazama A, Morishima S (1998) Criteria for the molecular identification of the volume-sensitive outwardly rectifying Cl^- channel. J Gen Physiol 112:365–367

Okada Y, Maeno E, Shimizu T, Dezaki K, Wang J, Morishima S (2001) Receptor-mediated control of regulatory volume decrease (RVD) and apoptotic volume decrease (AVD) J Physiol (Lond) 532:3–16

Oliet SH, Bourque CW (1993) Mechanosensitive channels transduce osmosensitivity in supraoptic neurons. Nature 364:341–343

Olson JE (1999) Osmolyte contents of cultured astrocytes grown in hypo-osmotic medium. Biochim Biophys Acta 1453:175–179

Ono S, Mougouris T, DuBose TD, Jr., Sansom SC (1994) ATP and calcium modulation of nonselective cation channels in IMCD cells. Am J Physiol 267:F558–F565

Orlic T, Loomis WH, Shreve A, Namiki S, Junger WG (2002) Hypertonicity increases cAMP in PMN and blocks oxidative burst by PKA-dependent and -independent mechanisms. Am J Physiol Cell Physiol 282:C1261–C1269

Orlowski J, Grinstein S (1997) Na^+/H^+ exchangers of mammalian cells. J Biol Chem 272:22373–22376

Pácha J, Frindt G, Sackin H, Palmer LG (1991) Apical maxi K channels in intercalated cells of CCT. Am J Physiol 261:F696–F705

Palmer CJ, Scott BT, Jones LR (1991) Purification and complete sequence determination of the major plasma membrane substrate for cAMP-dependent protein kinase and protein kinase C in myocardium. J Biol Chem 266:11126–11130

Papakonstanti EA, Vardaki EA, Stournaras C (2000) Actin cytoskeleton: a signaling sensor in cell volume regulation. Cell Physiol Biochem 10:257–264

Papassotiriou J, Eggermont J, Droogmans G, Nilius B (2001) Ca^{2+}-activated Cl^- channels in Ehrlich ascites tumor cells are distinct from mCLCA1:2 and 3. Pflugers Arch 442:273–279

Park K, Lee S, Elliott AC, Kim JS, Lee JH (2002) Swelling-induced Ca^{2+} release from intracellular calcium stores in rat submandibular gland acinar cells. J Membr Biol 186:165–176

Park K-P, Beck JS, Douglas IJ, Brown PD (1994) Ca^{2+}-activated K^+ channels are involved in regulatory volume decrease in acinar cells isolated from the rat lacrimal gland. J Membr Biol 141:193–201

Parker JC, Dunham PB, Minton AP (1995) Effects of ionic strength on the regulation of Na/H exchange and K-Cl cotransport in dog red blood cells. J Gen Physiol 105:677–699

Pasantes-Morales H, Chacon E, Sanchez-Olea R, Moran J (1994a) Volume regulation in cultured neurons: pivotal role of taurine. Adv Exp Med Biol 359:317–323

Pasantes-Morales H, Murray RA, Sanchez-Olea R, Moran J (1994b) Regulatory volume decrease in cultured astrocytes. II. Permeability pathway to amino acids and polyols. Am J Physiol 266:C172–C178

Pasantes-Morales H, Cardin V, Tuz K (2000a) Signaling events during swelling and regulatory volume decrease. Neurochem Res 25:1301–1314

Pasantes-Morales H, Franco R, Torres-Marquez ME, Hernandez-Fonseca K, Ortega A (2000b) Amino acid osmolytes in regulatory volume decrease and isovolumetric regulation in brain cells: contribution and mechanisms. Cell Physiol Biochem 10:361–370

Paulmichl M, Li Y, Wickman K, Ackerman M, Peralta E, Clapham D (1992) New mammalian chloride channel identified by expression cloning. Nature 356:238–241

Payne JA (1997) Functional characterization of the neuronal-specific K-Cl cotransporter: implications for $[K^+]_o$ regulation. Am J Physiol 273:C1516–C1525

Pearson MM, Lu J, Mount DB, Delpire E (2001) Localization of the K^+-Cl^- cotransporter, KCC3, in the central and peripheral nervous systems: expression in the choroid plexus, large neurons and white matter tracts. Neuroscience 103:481–491

Pedersen S, Hoffmann EK, Hougaard C, Lambert IH (2000) Cell shrinkage is essential in lysophosphatidic acid signaling in Ehrlich ascites tumor cells. J Membr Biol 173:19-29

Pedersen SF, Hoffmann EK, Mills JW (2001) The cytoskeleton and cell volume regulation. Comp Biochem Physiol A Mol Integr Physiol 130:385–399

Pedersen SF, Kramhoft B, Jorgensen NK, Hoffmann EK (1996) Shrinkage-induced activation of the Na^+/H^+ exchanger in Ehrlich ascites tumor cells: mechanisms involved in the activation and a role for the exchanger in cell volume regulation. J Membr Biol 149:141–159

Pedersen SF, Prenen J, Droogmans G, Hoffmann EK, Nilius B (1998) Separate swelling- and Ca^{2+}-activated anion currents in Ehrlich ascites tumor cells. J Membr Biol 163:97–110

Perlman DF, Goldstein L (1999) Organic osmolyte channels in cell volume regulation in vertebrates. J Exp Zool 283:725–733

Peters-Regehr T, Bode JG, Kubitz R, Häussinger D (1999) Organic osmolyte transport in quiescent and activated rat hepatic stellate cells (Ito cells) Hepatology 29:173–180

Petty CN, Lucero MT (1999) Characterization of a Na^+-dependent betaine transporter with Cl^- channel properties in squid motor neurons. J Neurophysiol 81:1567–1574

Pon DC, Hill CE (1997) Existence, properties, and functional expression of "Maxi- K"-type, Ca^{2+}-activated K^+ channels in short-term cultured hepatocytes. J Cell Physiol 171:87–94

Popp R, Hoyer J, Gögelein H (1993) Mechanosensitive nonselective cation channels in the antiluminal membrane of cerebral capillaries (blood-brain barrier) EXS 66:101–105

Popp R, Hoyer J, Meyer J, Galla H-J, Gögelein H (1992) Stretch-activated nonselective cation channels in the antiluminal membrane of porcine cerebral capillaries. J Physiol (Lond) 454:435–449

Porcellati F, Hosaka Y, Hlaing T, Togawa M, Larkin DD, Karihaloo A, Stevens MJ, Killen PD, Greene DA (1999) Alternate splicing in human Na^+-MI cotransporter gene yields differentially regulated transport isoforms. Am J Physiol 276:C1325–C1337

Preston AS, Yamauchi A, Kwon HM, Handler JS (1995) Activators of protein kinase A and of protein kinase C inhibit MDCK cell myo-inositol and betaine uptake. J Am Soc Nephrol 6:1559–1564

Qian X, Vinnakota S, Edwards C, Sarkar HK (2000) Molecular characterization of taurine transport in bovine aortic endothelial cells. Biochim Biophys Acta 1509:324–334

Race JE, Makhlouf FN, Logue PJ, Wilson FH, Dunham PB, Holtzman EJ (1999) Molecular cloning and functional characterization of KCC3 a new K-Cl cotransporter. Am J Physiol 277:C1210–C1219

Rasola A, Galietta LJ, Barone V, Romeo G, Bagnasco S (1995) Molecular cloning and functional characterization of a GABA/betaine transporter from human kidney. FEBS Lett 373:229–233

Ren XD, Kiosses WB, Schwartz MA (1999) Regulation of the small GTP-binding protein Rho by cell adhesion and the cytoskeleton. EMBO J 18:578–585

Renshaw MW, Toksoz D, Schwartz MA (1996) Involvement of the small GTPase rho in integrin-mediated activation of mitogen-activated protein kinase. J Biol Chem 271:21691–21694

Ritter M, Fürst J, Wöll E, Chwatal S, Gschwentner M, Lang F, Deetjen P, Paulmichl M (2001) Na^+/H^+ exchangers: Linking osmotic dysequilibrium to modified cell function. Cell Physiol Biochem 11:1–18

Ritter M, Steidl M, Lang F (1991) Inhibition of ion conductances by osmotic shrinkage of Madin-Darby canine kidney cells. Am J Physiol 261:C602–C607

Robson L, Hunter M (1994) Volume-activated, gadolinium-sensitive whole-cell currents in single proximal cells of frog kidney. Pflugers Arch 429:98–106

Roe MW, Moore AL, Lidofsky SD (2001) Purinergic-independent calcium signaling mediates recovery from hepatocellular swelling: implications for volume regulation. J Biol Chem 276:30871–30877

Roman R, Feranchak AP, Troetsch M, Dunkelberg JC, Kilic G, Schlenker T, Schaack J, Fitz JG (2002) Molecular characterization of volume-sensitive SK_{Ca} channels in human liver cell lines. Am J Physiol 282:G116–G122

Roman RM, Bodily KO, Wang Y, Raymond JR, Fitz JG (1998) Activation of protein kinase C alpha couples cell volume to membrane Cl^- permeability in HTC hepatoma and Mz-ChA-1 cholangiocarcinoma cells. Hepatology 28:1073–1080

Roman RM, Smith RL, Feranchak AP, Clayton GH, Doctor RB, Fitz JG (2001) ClC-2 chloride channels contribute to HTC cell volume homeostasis. Am J Physiol 280:G344–G353

Roy G (1995) Amino acid current through anion channels in cultured human glial cells. J Membr Biol 147:35–44

Roy G, Banderali U (1994) Channels for ions and amino acids in kidney cultured cells (MDCK) during volume regulation. J Exp Zool 268:121–126

Roy G, Malo C (1992) Activation of amino acid diffusion by a volume increase in cultured kidney (MDCK) cells. J Membr Biol 130:83–90

Ruhfus B, Kinne RKH (1996) Hypotonicity-activated efflux of taurine and myo-inositol in rat inner medullary collecting duct cells: evidence for a major common pathway. Kidney Blood Press Res 19:317–324

Ruhfus B, Tinel H, Kinne RKH (1996) Role of G-proteins in the regulation of organic osmolyte efflux from isolated rat renal inner medullary collecting duct cells. Pflugers Arch 433:35–41

Ruhfus B, Bauernschmitt HG, Kinne RKH (1998) Properties of a polarized primary culture from rat renal inner medullary collecting duct (IMCD) cells. In Vitro Cell Dev Biol Anim 34:227–231

Russell JM (2000) Sodium-potassium-chloride cotransport. Physiol Rev 80:211–276
Sackin H (1989) A stretch-activated K+ channel sensitive to cell volume. Proc Natl Acad Sci USA 86:1731–1735
Sadoshima J, Izumo S (1993) Mechanical stretch rapidly activates multiple signal transduction pathways in cardiac myocytes: potential involvement of an autocrine/paracrine mechanism. EMBO J 12:1681–1692
Saier MH, Jr (2000) Families of proteins forming transmembrane channels. J Membr Biol 175:165–180
Sajan MP, Bandyopadhyay G, Kanoh Y, Standaert ML, Quon MJ, Reed BC, Dikic I, Farese RV (2002) Sorbitol activates atypical protein kinase C and GLUT4 glucose transporter translocation/glucose transport through proline-rich tyrosine kinase-2, the extracellular signal-regulated kinase pathway and phospholipase D. Biochem J 362:665–674
Sakai H, Kakinoki B, Diener M, Takeguchi N (1996) Endogenous arachidonic acid inhibits hypotonically-activated Cl^- channels in isolated rat hepatocytes. Jpn J Physiol 46:311–318
Sanchez OR, Pasantes-Morales H, Lazaro A, Cereijido M (1991) Osmolarity-sensitive release of free amino acids from cultured kidney cells (MDCK) J Membr Biol 121:1–9
Sanchez-Olea R, Moran J, Schousboe A, Pasantes-Morales H (1991) Hypo-osmolarity-activated fluxes of taurine in astrocytes are mediated by diffusion. Neurosci Lett 130:233–236
Sanguinetti MC, Curran ME, Zou A, Shen J, Spector PS, Atkinson DL, Keating MT (1996) Coassembly of KvLQT1 and minK (IsK) proteins to form cardiac IKs potassium channel. Nature 384:80–83
Sardet C, Counillon L, Franchi A, Pouyssegur J (1990) Growth factors induce phosphorylation of the Na^+/H^+ antiporter, glycoprotein of 110 kD. Science 247:723–726
Sardet C, Fafournoux P, Pouyssegur J (1991) Alpha-thrombin, epidermal growth factor, and okadaic acid activate the Na^+/H^+ exchanger, NHE-1, by phosphorylating a set of common sites. J Biol Chem 266:19166–19171
Satsu H, Miyamoto Y, Shimizu M (1999) Hypertonicity stimulates taurine uptake and transporter gene expression in Caco-2 cells. Biochim Biophys Acta 1419:89–96
Schirmer T, Keller TA, Wang YF, Rosenbusch JP (1995) Structural basis for sugar translocation through maltoporin channels at 3.1 A resolution. Science 267:512–514
Schlatter E (1993) Regulation of ion channels in the cortical collecting duct. Renal Physiol Biochem 16:21–36
Schlatter E, Ankorina-Stark I, Cermak R, Haxelmans S, Kleta R, Hirsch JR (1997) Cell shrinkage activates a cation conductance in principal cells of rat cortical collecting duct. Cell Physiol Biochem 7:321-332
Schliess F, Sinning R, Fischer R, Schmalenbach C, Häussinger D (1996) Calcium-dependent activation of Erk-1 and Erk-2 after hypo-osmotic astrocyte swelling. Biochem J 320:167–171
Schmidt-Rose T, Jentsch TJ (1997) Transmembrane topology of a CLC chloride channel. Proc Natl Acad Sci USA 94:7633–7638
Schneider SW, Pagel P, Rotsch C, Danker T, Oberleithner H, Radmacher M, Schwab A (2000) Volume dynamics in migrating epithelial cells measured with atomic force microscopy. Pflugers Arch 439:297–303
Schoenwaelder SM, Burridge K (1999) Bidirectional signaling between the cytoskeleton and integrins. Curr Opin Cell Biol 11:274–286
Schreiber M, Salkoff L (1997) A novel calcium-sensing domain in the BK channel. Biophys J 73:1355–1363
Schumacher PA, Sakellaropoulos G, Phipps DJ, Schlichter LC (1995) Small-conductance chloride channels in human peripheral T lymphocytes. J Membr Biol 145:217–232
Schütt W, Sackin H (1997) A new technique for evaluating volume sensitivity of ion channels. Pflugers Arch 433:368–375
Schwab A, Oberleithner H (1996) Plasticity of renal epithelial cells: The way a potassium channel supports migration. Pflugers Arch 432:R87-R93
Schwab A, Westphale H-J, Wojnowski L, Wünsch S, Oberleithner H (1993) Spontaneously oscillating K^+ channel activity in transformed Madin-Darby canine kidney cells. J Clin Invest 92:218–223
Schwab A, Schuricht B, Seeger P, Reinhardt J, Dartsch PC (1999) Migration of transformed renal epithelial cells is regulated by K^+ channel modulation of actin cytoskeleton and cell volume. Pflugers Arch 438:330–337
Schwiebert EM, Mills JW, Stanton BA (1994) Actin-based cytoskeleton regulates a chloride channel and cell volume in a renal cortical collecting duct cell line. J Biol Chem 269:7081–7089
Setiawan I, Henke G, Feng Y, Böhmer C, Vasilets LA, Schwarz W, Lang F (2002) Stimulation of *Xenopus* oocyte Na^+,K^+ ATPase by the serum and glucocorticoid-dependent kinase sgk1. Pflugers Arch 444:426–431

Shen MR, Chou CY, Browning JA, Wilkins RJ, Ellory JC (2001) Human cervical cancer cells use Ca^{2+} signalling, protein tyrosine phosphorylation and MAP kinase in regulatory volume decrease. J Physiol 537:347–362

Shen MR, Furla P, Chou CY, Ellory JC (2002) Myosin light chain kinase modulates hypotonicity-induced Ca^{2+} entry and Cl^- channel activity in human cervical cancer cells. Pflugers Arch 444:276–285

Shennan D, McNeillie S, Curran D (1994) The effect of a hypo-osmotic shock on amino acid efflux from lactating rat mammary tissue: stimulation of taurine and glycine efflux via a pathway distinct from anion exchange and volume-activated anion channels. Exp Physiol 79:797–808.

Shiga N, Wangemann P (1995) Ion selectivity of volume regulatory mechanisms present during a hypo-osmotic challenge in vestibular dark cells. Biochim Biophys Acta 1240:48–54

Shinozuka K, Tanaka N, Kawasaki K, Mizuno H, Kubota Y, Nakamura K, Hashimoto M, Kunitomo M (2001) Participation of ATP in cell volume regulation in the endothelium after hypotonic stress. Clin Exp Pharmacol Physiol 28:799–803

Shrode LD, Krump E, Grinstein S (1998) Activation of protein kinases upon volume changes: role in cellular homeostasis. Contrib Nephrol 123:79–93

Shrode LD, Rubie EA, Woodgett JR, Grinstein S (1997) Cytosolic alkalinization increases stress-activated protein kinase/c-Jun NH2-terminal kinase (SAPK/JNK) activity and p38 mitogen-activated protein kinase activity by a calcium-independent mechanism J Biol Chem 272:13653–13659

Siebens AW, Spring KR (1989) A novel sorbitol transport mechanism in cultured renal papillary epithelial cells. Am J Physiol 257:F937–F946

Sizeland PC, Chambers ST, Lever M, Bason LM, Robson RA (1993) Organic osmolytes in human and other mammalian kidneys. Kidney Int 43:448–453

Skou JC (1957) The influence of some cations on an adenosine triphosphatase from peripheral nerves. Biochim Biophys Acta 23:394–401

Sokabe M, Sachs F, Jing ZQ (1991) Quantitative video microscopy of patch clamped membranes stress, strain, capacitance, and stretch channel activation. Biophys J 59:722–728

Soleimani M, Singh G, Bizal GL, Gullans SR, McAteer JA (1994) Na^+/H^+ exchanger isoforms NHE-2 and NHE-1 in inner medullary collecting duct cells. Expression, functional localization, and differential regulation. J Biol Chem 269:27973–27978

Staines HM, Godfrey EM, Lapaix F, Egee S, Thomas S, Ellory JC (2002) Two functionally distinct organic osmolyte pathways in Plasmodium gallinaceum-infected chicken red blood cells. Biochim Biophys Acta 1561:98–108

Stegen C, Matskevich I, Wagner CA, Paulmichl M, Lang F, Broer S (2000) Swelling-induced taurine release without chloride channel activity in Xenopus laevis oocytes expressing anion channels and transporters. Biochim Biophys Acta 1467:91–100

Steidl M, Ritter M, Lang F (1991) Regulation of potassium conductance by prostaglandins in cultured renal epitheloid (Madin-Darby canine kidney) cells. Pflugers Arch 418:431–436

Stephens L, Anderson K, Stokoe D, Erdjument-Bromage H, Painter GF, Holmes AB, Gaffney PR, Reese CB, McCormick F, Tempst P, Coadwell J, Hawkins PT (1998) Protein kinase B kinases that mediate phosphatidylinositol 3,4,5-trisphosphate-dependent activation of protein kinase B. Science 279:710–714

Stoddard JS, Steinbach JH, Simchowitz L (1993) Whole cell Cl^- currents in human neutrophils induced by cell swelling. Am J Physiol 265:C156–C165

Stoner LC, Morley GE (1995) Effect of basolateral or apical hypo-osmolarity on apical maxi K channels of everted rat collecting tubule. Am J Physiol 268:F569–F580

Strange K (1992) Regulation of solute and water balance and cell volume in the central nervous system. J Am Soc Nephrol 3:12–27

Strange K (1998) Molecular identity of the outwardly rectifying, swelling-activated anion channel: Time to reevaluate pICln. J Gen Physiol 111:617–622

Strange K, Emma F, Jackson PS (1996) Cellular and molecular physiology of volume-sensitive anion channels. Am J Physiol 270:C711–C730

Strange K, Morrison R, Heilig CW, DiPietro S, Gullans SR (1991) Upregulation of inositol transport mediates inositol accumulation in hyperosmolar brain cells. Am J Physiol 260:C784–C790

Strotmann R, Harteneck C, Nunnenmacher K, Schultz G, Plant TD (2000) OTRPC4 a nonselective cation channel that confers sensitivity to extracellular osmolarity. Nat Cell Biol 2:695–702

Stühmer W, Ruppersberg JP, Schroter KH, Sakmann B, Stocker M, Giese KP, Perschke A, Baumann A, Pongs O (1989) Molecular basis of functional diversity of voltage-gated potassium channels in mammalian brain. EMBO J 8:3235–3244

Stutzin A, Torres R, Oporto M, Pacheco P, Eguiguren AL, Cid LP, Sepulveda FV (1999) Separate taurine and chloride efflux pathways activated during regulatory volume decrease. Am J Physiol 277:C392–C402

Su W, Shmukler BE, Chernova MN, Stuart-Tilley AK, De Franceschi L, Brugnara C, Alper SL (1999) Mouse K-Cl cotransporter KCC1: cloning, mapping, pathological expression, and functional regulation. Am J Physiol 277:C899–C912

Sun AM, Liu Y, Dworkin LD, Tse CM, Donowitz M, Yip KP (1997) Na^+/H^+ exchanger isoform 2 (NHE2) is expressed in the apical membrane of the medullary thick ascending limb. J Membr Biol 160:85–90

Suzuki M, Kawahara K, Ogawa A, Morita T, Kawaguchi Y, Kurihara S, Sakai O (1990) $[Ca^{2+}]_i$ rises via G protein during regulatory volume decrease in rabbit proximal tubule cells. Am J Physiol 258:F690–F696

Suzuki M, Sato J, Kutsuwada K, Ooki G, Imai M (1999) Cloning of a stretch-inhibitable nonselective cation channel. J Biol Chem 274:6330–6335

Szaszi K, Grinstein S, Orlowski J, Kapus A (2000) Regulation of the epithelial Na^+ /H^+ exchanger isoform by the cytoskeleton. Cell Physiol Biochem 10:265–272

Szászi K, Kurashima K, Kapus A, Paulsen A, Kaibuchi K, Grinstein S, Orlowski J (2000) RhoA and rho kinase regulate the epithelial Na^+/H^+ exchanger NHE3 - Role of myosin light chain phosphorylation. J Biol Chem 275:28599–28606

Taglietti V, Toselli M (1988) A study of stretch-activated channels in the membrane of frog oocytes: Interaction with Ca^{2+} ions. J Physiol (Lond) 407:311–328

Tai KK, Wang KW, Goldstein SA, N (1997) MinK potassium channels are heteromultimeric complexes. J Biol Chem 272:1654–1658

Takenaka M, Bagnasco SM, Preston AS, Uchida S, Yamauchi A, Kwon HM, Handler JS (1995) The canine betaine gamma-amino-n-butyric acid transporter gene: diverse mRNA isoforms are regulated by hypertonicity and are expressed in a tissue-specific manner. Proc Natl Acad Sci USA 92:1072–1076

Takumi T, Ohkubo H, Nakanishi S (1989) Cloning of a membrane protein that induces a slow voltage-gated potassium current. Science 242:1042–1045

Tamkun MM, Knoth KM, Walbridge JA, Kroemer H, Roden DM, Glover DM (1991) Molecular cloning and characterization of two voltage-gated K^+ channel cDNAs from human ventricle. FASEB J 5:331–337

Taniguchi J, Guggino WB (1989) Membrane stretch: A physiological stimulator of Ca^{2+}-activated K^+ channels in thick ascending limb. Am J Physiol 257:F347–F352

Terada Y, Inoshita S, Hanada S, Shimamura H, Kuwahara M, Ogawa W, Kasuga M, Sasaki S, Marumo F (2001) Hyperosmolality activates Akt and regulates apoptosis in renal tubular cells. Kidney Int 60:553–567

Tewari KP, Malinowska DH, Sherry AM, Cuppoletti J (2000) PKA and arachidonic acid activation of human recombinant ClC-2 chloride channels. Am J Physiol Cell Physiol 279:C40–C50

Theodoropoulos PA, Stournaras C, Stoll B, Markogiannakis E, Lang F, Gravanis A, Häussinger D (1992) Hepatocyte swelling leads to rapid decrease of the G-/total actin ratio and increases actin mRNA levels. FEBS Lett 311:241–245

Therien AG, Blostein R (2000) Mechanisms of sodium pump regulation. Am J Physiol Cell Physiol 279:C541–C566

Thinnes FP, Walter G, Hellmann KP, Hellmann T, Merker R, Kiafard Z, Eben-Brunnen J, Schwarzer C, Gotz H, Hilschmann N (2001) Gadolinium as an opener of the outwardly rectifying Cl^- channel (ORCC) Is there relevance for cystic fibrosis therapy? Pflugers Arch 443:S111-S116

Thoroed SM, Lauritzen L, Lambert IH, Hansen HS, Hoffmann EK (1997) Cell swelling activates phospholipase A2 in Ehrlich ascites tumor cells. J Membr Biol 160:47–58

Tilly BC, van den Berghe N, Tertoolen LG, Edixhoven MJ, de Jonge HR (1993) Protein tyrosine phosphorylation is involved in osmoregulation of ionic conductances. J Biol Chem 268:19919–19922

Tinel H, Petersen OH (2000) Transport across intracellular organelles. In Seldin DW, Giebisch G (eds) The kidney. Physiology and pathophysiology. Lippincott Williams and Wilkins, Philadelphia, pp 294–319

Tinel H, Wehner F, Sauer H (1994) Intracellular Ca^{2+} release and Ca^{2+} influx during regulatory volume decrease in IMCD cells. Am J Physiol 267:F130–F138

Tinel H, Wehner F, Kinne RKH (1997) Arachidonic acid as a second messenger for hypotonicity-induced calcium transients in rat IMCD cells. Pflugers Arch 433:245–253

Tinel H, Cancela JM, Mogami H, Gerasimenko JV, Gerasimenko OV, Tepikin AV, Petersen OH (1999) Active mitochondria surrounding the pancreatic acinar granule region prevent spreading of inositol trisphosphate-evoked local cytosolic Ca^{2+} signals. EMBO J 18:4999–5008

Tinel H, Kinne-Saffran E, Kinne RKH (2000) Calcium signalling during RVD of kidney cells. Cell Physiol Biochem 10:297–302

Tinel H, Kinne-Saffran E, Kinne RKH (2002) Calcium-induced calcium release participates in cell volume regulation of rabbit TALH cells. Pflugers Arch 443:754–761

Tominaga T, Barber DL (1998) Na-H exchange acts downstream of RhoA to regulate integrin-induced cell adhesion and spreading. Mol Biol Cell 9:2287–2303

Tominaga T, Ishizaki T, Narumiya S, Barber DL (1998) p160ROCK mediates RhoA activation of Na-H exchange. EMBO J 17:4712–4722

Toro L, Wallner M, Meera P, Tanaka Y (1998) Maxi-K_{Ca} a unique member of the voltage-gated K channel superfamily. News Physiol Sci 13:112–117

Toyomoto T, Knutsen D, Soos G, Sato K (1997) Na-K-2Cl cotransporters are present and regulated in simian eccrine clear cells. Am J Physiol 273:R270–R277

Tse C-M, Levine SA, Yun CHC, Brant SR, Nath S, Pouysségur J, Donowitz M (1994) Molecular properties, kinetics and regulation of mammalian Na^+/H^+ exchangers. Cell Physiol Biochem 4:282–300

Tsuganezawa H, Kobayashi K, Iyori M, Araki T, Koizumi A, Watanabe S, Kaneko A, Fukao T, Monkawa T, Yoshida T, Kim DK, Kanai Y, Endou H, Hayashi M, Saruta T (2001) A new member of the HCO_3^- transporter superfamily is an apical anion exchanger of α-intercalated cells in the kidney. J Biol Chem 276:8180–8189

Tsumura T, Oiki S, Ueda S, Okuma M, Okada Y (1996) Sensitivity of volume-sensitive Cl^- conductance in human epithelial cells to extracellular nucleotides. Am J Physiol 271:C1872–C1878

Ubl J, Murer H, Kolb HA (1988) Hypotonic shock evokes opening of Ca^{2+}-activated K channels in opossum kidney cells. Pflugers Arch 412:551–553

Uchida S, Nakanishi T, Kwon HM, Preston AS, Handler JS (1991) Taurine behaves as an osmolyte in Madin-Darby canine kidney cells. Protection by polarized, regulated transport of taurine. J Clin Invest 88:656–662

Uchida S, Kwon HM, Yamauchi A, Preston AS, Marumo F, Handler JS (1992) Molecular cloning of the cDNA for an MDCK cell Na^+- and Cl^--dependent taurine transporter that is regulated by hypertonicity. Proc Natl Acad Sci USA 89:8230–8234

Urbach V, Leguen I, O'Kelly I, Harvey BJ (1999) Mechanosensitive calcium entry and mobilization in renal A6 cells. J Membr Biol 168:29–37

Valverde MA, Díaz M, Sepúlveda FV, Gill DR, Hyde SC, Higgins CF (1992) Volume-regulated chloride channels associated with the human multidrug-resistance P-glycoprotein. Nature 355:830–833

Van der Kaay J, Beck M, Gray A, Downes CP (1999) Distinct phosphatidylinositol 3-kinase lipid products accumulate upon oxidative and osmotic stress and lead to different cellular responses. J Biol Chem 274:35963–35968

van Der Wijk T, Dorrestijn J, Narumiya S, Maassen JA, de Jonge HR, Tilly BC (1998) Osmotic swelling-induced activation of the extracellular-signal-regulated protein kinases Erk-1 and Erk-2 in intestine 407 cells involves the Ras/Raf-signalling pathway. Biochem J 331:863–869

van Der Wijk T, Tomassen SF, de Jonge HR, Tilly BC (2000) Signalling mechanisms involved in volume regulation of intestinal epithelial cells. Cell Physiol Biochem 10:289–296

van Driessche W, Erlij D (1994) Cell swelling activates a poorly selective monovalent cation channel in the apical membrane of toad urinary bladder. Pflugers Arch 428:1–8

van Driessche W, De Smet P, De Smedt H (1994) Poorly selective cation channels in the apical membrane of A6 cells. Pflugers Arch 426:387–395

Vandorpe DH, Shmukler BE, Jiang L, Lim B, Maylie J, Adelman JP, De Franceschi L, Capellini MD, Brugnara C, Alper SL (2002) cDNA cloning and functional characterization of the moese Ca^{2+}-gated K^+ channel, mIK1. J Biol Chem 273:21542–21553

Vanoye CG, Reuss L (1999) Stretch-activated single K^- channels account for whole-cell currents elicited by swelling. Proc Natl Acad Sci USA 96:6511–6516

Verbalis JG (1994) Pathogenesis of hyponatremia in an experimental model of the syndrome of inappropriate antidiuresis. Am J Physiol 267:R1617–R1625

Verbalis JG, Drutarosky MD, Ertel RJ, Vollmer RR (1989) Adaptive responses to sustained volume expansion in hyponatraemic rats. J Endocrinol 122:421–431

Vergara C, Latorre R, Marrion NV, Adelman JP (1998) Calcium-activated potassium channels. Curr Opin Neurobiol 8:321–329

Vexler ZS, Symons M, Barber DL (1996) Activation of Na^+-H^+ exchange is necessary for RhoA-induced stress fiber formation. J Biol Chem 271:22281–22284

Viana F, de la PE, Pecson B, Schmidt RF, Belmonte C (2001) Swelling-activated calcium signalling in cultured mouse primary sensory neurons. Eur J Neurosci 13:722–734

Vieira LL, Lafuente E, Blum J, Cabantchik ZI (1997) Modulation of the swelling-activated amino acid channel of Leishmania major promastigotes by protein kinases. Mol Biochem Parasitol 90:449–461

Volk T, Frömter E, Korbmacher C (1995) Hypertonicity activates nonselective cation channels in mouse cortical collecting duct cells. Proc Natl Acad Sci USA 92:8478–8482

Völkl H, Lang F (2001) Electrophysiology of betaine transport in isolated perfused straight proximal tubule. Pflugers Arch 442:136–140

Vom Dahl S, Hallbrucker C, Lang F, Häussinger D (1991) Role of eicosanoids, inositol phosphates and extracellular Ca^{2+} in cell-volume regulation of rat liver. Eur J Biochem 198:73–83

Vom Dahl S, Dombrowski F, Schmitt M, Schliess F, Pfeifer U, Häussinger D (2001) Cell hydration controls autophagosome formation in rat liver in a microtubule-dependent way downstream from p38MAPK activation. Biochem J 354:31–36

von Weikersthal SF, Hickman ME, Hladky SB, Barrand MA (1997) Hypotonicity-induced changes in anion permeability of cultured rat brain endothelial cells. Biochim Biophys Acta 1325:99–107

Wagner CA, Ott M, Klingel K, Beck S, Melzig J, Friedrich B, Wild KN, Bröer S, Moschen I, Albers A, Waldegger S, Tummler B, Egan ME, Geibel JP, Kandolf R, Lang F (2001) Effects of the serine/threonine kinase SGK1 on the epithelial Na^+ channel (ENaC) and CFTR: implications for cystic fibrosis. Cell Physiol Biochem 11 209–218

Walaas SI, Czernik AJ, Olstad OK, Sletten K, Walaas O (1994) Protein kinase C and cyclic AMP-dependent protein kinase phosphorylate phospholemman, an insulin and adrenaline- regulated membrane phosphoprotein, at specific sites in the carboxy terminal domain. Biochem J 304:635–640

Waldegger S, Barth P, Raber G, Lang F (1997) Cloning and characterization of a putative human serine/threonine protein kinase transcriptionally modified during anisotonic and isotonic alterations of cell volume. Proc Natl Acad Sci USA 94:4440–4445

Wang K, Wondergem R (1991) Effects of hyperosmotic medium on hepatocyte volume, transmembrane potential and intracellular K^+ activity. Biochim Biophys Acta 1069:187–196

Wang Y, Roman R, Lidofsky SD, Fitz JG (1996) Autocrine signaling through ATP release represents a novel mechanism for cell volume regulation. Proc Natl Acad Sci USA 93:12020–12025

Wangemann P, Liu J, Shen Z, Shipley A, Marcus DC (1995) Hypo-osmotic challenge stimulates transepithelial K^+ secretion and activates apical IsK channel in vestibular dark cells. J Membr Biol 147:263–273

Wärntges S, Friedrich B, Henke G, Duranton C, Lang PA, Waldegger S, Meyermann R, Kuhl D, Speckmann EJ, Obermüller N, Witzgall R, Mack AF, Wagner HJ, Wagner A, Bröer S, Lang F (2002) Cerebral localization and regulation of the cell volume-sensitive serum- and glucocorticoid-dependent kinase SGK1. Pflugers Arch 443:617–624

Warskulat U, Wettstein M, Häussinger D (1997a) Osmo-regulated taurine transport in H4IIE hepatoma cells and perfused rat liver. Biochem J 321:83–690

Warskulat U, Zhang F, Häussinger D (1997b) Taurine is an osmolyte in rat liver macrophages (Kupffer cells) J Hepatol 26:1340–1347

Warth R, Hamm K, Bleich M, Kunzelmann E, Von Hahn T, Schreiber R, Ullrich E, Mengel M, Trautmann N, Kindle P, Schwab A, Greger R (1999) Molecular and functional characterization of the small Ca^{2+}-regulated K^+ channel (rSK4) of colonic crypts. Pflugers Arch 438:437–444

Watson AJ, Levine S, Donowitz M, Montrose MH (1992) Serum regulates Na^+/H^+ exchange in Caco-2 cells by a mechanism which is dependent on F-actin. J Biol Chem 267:956–962

Watson PA, Giger KE, Frankenfield CM (1991) Activation of adenylate cyclase during swelling of S49 cells in hypotonic medium is not involved in subsequent volume regulation. Mol Cell Biochem 104:51–56

Weaver YR, Kiessling K, Cossins AR (1999) Responses of the Na^+/H^+ exchanger of European flounder red blood cells to hypertonic, β-adrenergic and acidotic stimuli. J Exp Biol 202:21–32

Weekes J, Barry ST, Critchley DR (1996) Acidic phospholipids inhibit the intramolecular association between the N- and C-terminal regions of vinculin, exposing actin-binding and protein kinase C phosphorylation sites. Biochem J 314:827–832

Wehner F (1998) Cell volume-regulated cation channels. Contrib Nephrol 123:8–20

Wehner F, Tinel H (1998) Role of Na^+ conductance, Na^+-H^+ exchange, and Na^+- K^+-$2Cl^-$ symport in the regulatory volume increase of rat hepatocytes. J Physiol (Lond) 506:127–142

Wehner F, Tinel H (2000) Osmolyte and Na^+ transport balances of rat hepatocytes as a function of hypertonic stress. Pflugers Arch 441:12–24

Wehner F, Sauer H, Kinne RKH (1995) Hypertonic stress increases the Na^+ conductance of rat hepatocytes in primary culture. J Gen Physiol 105:507–535

Wehner F, Tinel H, Kinne RKH (1997) Pharmacology of volume activated Na^+ conductance in rat hepatocytes. Physiologist 40:A-4

Wehner F, Lawonn P, Tinel H (2002a) Ionic mechanisms of regulatory volume increase (RVI) in the human hepatoma cell-line HepG2. Pflugers Arch 443:779–790

Wehner F, Olsen H, Bierhals K, Lin C-T, Waldmann H (2002b) Signal transduction in the hypertonic activation of rat hepatocyte Na^+ conductance. FASEB J A56

Weik C, Warskulat U, Bode J, Peters-Regehr T, Häussinger D (1998) Compatible organic osmolytes in rat liver sinusoidal endothelial cells. Hepatology 27:569–575

Weiss H, Lang F (1992) Ion channels activated by swelling of Madin Darby canine kidney (MDCK) cells. J Membr Biol 126:109–114

Weyand B, Warth R, Bleich M, Kerstan D, Nitschke R, Greger R (1998) Hypertonic cell shrinkage reduces the K^+ conductance of rat colonic crypts. Pflugers Arch 436:227–232

Willumsen NJ, Davis CW, Boucher RC (1994) Selective response of human airway epithelia to luminal but not serosal solution hypertonicity. Possible role for proximal airway epithelia as an osmolality transducer. J Clin Invest 94:779–787

Winpenny JP, Mathews CJ, Verdon B, Wardle CJ, C., Chambers JA, Harris A, Argent BE, Gray MA (1996) Volume-sensitive chloride currents in primary cultures of human fetal vas deferens epithelial cells. Pflugers Arch 432:644–654

Wirthensohn G, Lefrank S, Schmolke M, Guder WG (1989) Regulation of organic osmolyte concentrations in tubules from rat renal inner medulla. Am J Physiol 256:F128–F135

Wissenbach U, Bodding M, Freichel M, Flockerzi V (2000) Trp12 a novel Trp related protein from kidney. FEBS Lett 485:127–134

Wittels KA, Hubert EM, Musch MW, Goldstein L (2000) Osmolyte channel regulation by ionic strength in skate RBC. Am J Physiol Regul Integr Comp Physiol 279:R69–R76

Wojnowski L, Oberleithner H (1991) Hypertonicity in fused Madin-Darby canine kidney cells: transient rise in $NaHCO_3$ followed by sustained KCl accumulation. Pflugers Arch 419:43–50

Wolff NA, Kinne R (1988) Taurine transport by rabbit kidney brush-border membranes: coupling to sodium, chloride, and the membrane potential. J Membr Biol 102:131–139

Wright EM, Diamond JM (1977) Anion selectivity in biological systems. Physiol Rev 57:109–156

Wu X, Yang H, Iserovich P, Fischbarg J, Reinach PS (1997) Regulatory volume decrease by SV40-transformed rabbit corneal epithelial cells requires ryanodine-sensitive Ca^{2+}-induced Ca^{2+} release. J Membr Biol 158:127–136

Wymann MP, Pirola L (1998) Structure and function of phosphoinositide 3-kinases. Biochim Biophys Acta 1436:127–150

Xiong H, Li C, Garami E, Wang Y, Ramjeesingh M, Galley K, Bear CE (1999) ClC-2 activation modulates regulatory volume decrease. J Membr Biol 167:215–221

Xu B, Lu L (1994) Protein kinase A-regulated Cl^- channel in ML-1 human hematopoietic myeloblasts. J Membr Biol 142:65–75

Xu WX, Kim SJ, So I, Kang TM, Rhee JC, Kim KW (1997) Volume-sensitive chloride current activated by hypo-osmotic swelling in antral gastric myocytes of the guinea-pig. Pflugers Arch 435:9–19

Yamaguchi DT, Green J, Kleeman CR, Muallem S (1989) Characterization of volume-sensitive, calcium-permeating pathways in the osteosarcoma cell line UMR-106–01. J Biol Chem 264:4383–4390

Yamamoto Y, Suzuki H (1996) Two types of stretch-activated channel activities in guinea-pig gastric smooth muscle cells. Jpn J Physiol 46:337–345

Yamashita T, Yamauchi A, Miyai A, Taniguchi M, Yoshimine T, Tohyama M (1999) Neuroprotective role of Na^+/myo-inositol cotransporter against veratridine cytotoxicity. J Neurochem 72:1864–1870

Yamauchi A, Kwon HM, Uchida S, Preston AS, Handler JS (1991) Myo-inositol and betaine transporters regulated by tonicity are basolateral in MDCK cells. Am J Physiol 261:F197–F202

Yamauchi A, Uchida S, Kwon HM, Preston AS, Robey RB, Garcia-Perez A, Burg MB, Handler JS (1992) Cloning of a Na^+- and Cl^--dependent betaine transporter that is regulated by hypertonicity. J Biol Chem 267:649–652

Yamauchi A, Uchida S, Preston AS, Kwon HM, Handler JS (1993) Hypertonicity stimulates transcription of gene for Na^+-myo-inositol cotransporter in MDCK cells. Am J Physiol 264:F20–F23

Yamauchi A, Miyai A, Shimada S, Minami Y, Tohyama M, Imai E, Kamada T, Ueda N (1995) Localization and rapid regulation of Na^+/myo-inositol cotransporter in rat kidney. J Clin Invest 96:1195–1201

Yancey PH, Burg MB (1989) Distribution of major organic osmolytes in rabbit kidneys in diuresis and antidiuresis. Am J Physiol 257:F602–F607

Yancey PH, Clark ME, Hand SC, Bowlus RD, Somero GN (1982) Living with water stress: evolution of osmolyte systems. Science 217:1214–1222

Yang X-C, Sachs F (1989) Block of stretch-activated ion channels in Xenopus oocytes by gadolinium and calcium ions. Science 243:1068–1071

Yip KP (2002) Coupling of vasopressin-induced intracellular Ca^{2+} mobilization and apical exocytosis in perfused rat kidney collecting duct. J Physiol 538:891–899

Yordy MR, Bowen JW (1993) Na,K-ATPase expression and cell volume during hypertonic stress in human renal cells. Kidney Int 43:940–948

Yorek MA, Dunlap JA, Lowe WL, Jr (1999) Osmotic regulation of the Na^+/myo-inositol cotransporter and postinduction normalization. Kidney Int 55:215–224

Yu W-G, Sokabe M (1997) Hypotonically induced whole-cell currents in A6 cells: relationship with cell volume and cytoplasmic Ca^{2+}. Jpn J Physiol 47:553–565

Zablocki K, Miller SP, Garcia-Perez A, Burg MB (1991) Accumulation of glycerophosphocholine (GPC) by renal cells: osmotic regulation of GPC:choline phosphodiesterase. Proc Natl Acad Sci USA 88:7820–7824

Zelikovic I, Stejskal-Lorenz E, Lohstroh P, Budreau A, Chesney RW (1989) Anion dependence of taurine transport by rat renal brush-border membrane vesicles. Am J Physiol 256:F646–F655

Zelikovic I, Budreau-Patters A (1999) Cl^- and membrane potential dependence of amino acid transport across the rat renal brush border membrane. Mol Genet Metab 67:236–247

Zhang JJ, Jacob TJ, C (1997) Three different Cl^- channels in the bovine ciliary epithelium activated by hypotonic stress. J Physiol (Lond) 499:379–389

Zhang MI, O'Neil RG (1996) An L-type calcium channel in renal epithelial cells. J Membr Biol 154:259–266

Zhang Z, Cohen DM (1996) NaCl but not urea activates p38 and jun kinase in mIMCD3 murine inner medullary cells. Am J Physiol 271:F1234–F1238

Zhou C, Agarwal N, Cammarata PR (1994) Osmoregulatory alterations in myo-inositol uptake by bovine lens epithelial cells. Part 2: Cloning of a 626 bp cDNA portion of a Na^+/myo-inositol cotransporter, an osmotic shock protein. Invest Ophthalmol Vis Sci 35:1236–1242

Zhou C, Cammarata PR (1999) Characterization of the minimal osmotic response element of the bovine Na^+/myo-inositol cotransporter gene. Exp Eye Res 68:137–142

Zhuang S, Hirai SI, Ohno S (2000) Hyperosmolality induces activation of cPKC and nPKC, a requirement for ERK1/2 activation in NIH/3T3 cells. Am J Physiol Cell Physiol 278:C102–C109

Rev Physiol Biochem Pharmacol (2003) 148:81–104
DOI 10.1007/s10254-003-0010-4

E. Krüger · U. Kuckelkorn · A. Sijts · P.-M. Kloetzel

The components of the proteasome system and their role in MHC class I antigen processing

Published online: 25 March 2003

Abstract By generating peptides from intracellular antigens which are then presented to T cells, the ubiquitin/26S proteasome system plays a central role in the cellular immune response. The proteolytic properties of the proteasome are adapted to the requirements of the immune system by proteasome components whose synthesis is under the control of interferon-γ. Among these are three subunits with catalytic sites that are incorporated into the enzyme complex during its de novo synthesis. Thus, the proteasome assembly pathway and the formation of immunoproteasomes play a critical regulatory role in the regulation of the proteasome's catalytic properties. In addition, interferon-γ also induces the synthesis of the proteasome activator PA28 which, as part of the so-called hybrid proteasome, exerts a more selective function in antigen presentation. Consequently, the combination of a number of regulatory events tunes the proteasome system to gain maximal efficiency in the generation of peptides with regard to their quality and quantity.

The proteasome system is the central proteolytic system of the eukaryotic cell (Rock et al. 1994). The generation of MHC class I peptides by the proteasome is only one aspect of its various proteolytic functions. The major functions of the proteasome are the selective ATP-dependent degradation of cytosolic, nuclear, and membrane bound proteins (Plemper and Wolf 1999), as well as the regulation of such important cellular processes as cell cycle progression, transcription, differentiation, apoptosis, and stress response (Kloetzel 2001; Rock et al. 2002; Glickman et al. 2002).

The enzyme complex responsible for the selectivity of intracellular protein degradation is the 26S proteasome that degrades polyubiquitinated substrates or in some cases even unmodified proteins in an ATP-dependent manner. The 26S proteasome complex is formed by the 20S core particle harboring the proteolytically active sites and two 19S regulatory particles that are responsible for substrate interaction.

E. Krüger · U. Kuckelkorn · A. Sijts · P.-M. Kloetzel (✉)
Institut für Biochemie, Medizinische Fakultät, Humboldt-Universität zu Berlin, Charité, Monbijoustr. 2, 10117 Berlin, Germany
e-mail: p-m.kloetzel@charite.de

The 20S proteasome

One important prerequisite for the selective degradation of proteins resides in the structure and the specific catalytic features of the proteasome complex. The 20S catalytic core is a barrel-shaped particle composed four heptameric rings. The two outer rings consist of seven different but homologous α subunits (α1–α7), which provide the proteasome structure, control the access to the catalytic chamber, and interact with regulatory factors and complexes (Groll et al. 1997). The two inner rings are each composed of seven β subunits (β1–β7), three of which (i.e., β1-δ, β2-z, and β5-MB1) display catalytic activity.

The two α rings and the two β rings form two so-called antechambers through which substrates have to pass in order to reach the active site centers that are oriented towards the proteasomal lumen. The central gate formed by the α subunits is normally closed by the N-termini of the seven α subunits which keep the proteasome in a so-called latent, proteolytically inactive state (Groll et al. 1997; Groll et al. 2000). Thereby, the N-terminus of the α3 subunit interacts with conserved residues on the other α subunits (Köhler et al. 2001). Deletion of the N-terminus or the deletion of Asp 9 in the α3 subunit results in the opening of the gate and the facilitated access of substrates, a process also called gating. It is, however, important to note that even when the gate is open polypeptide substrates have to be unfolded to gain access to the catalytic cavity of the 20S core complex (Groll et al. 1997).

Due to its specific catalytic mechanisms that separate the proteasome from the four classical families of proteases, the proteasome belongs to the family of N-terminal hydrolases as the three catalytic β subunits contain an amino-terminal threonine residue whose hydroxyl group serves as nucleophile that attacks the peptide bond (Brannigan et al. 1995; Löwe et al. 1995; Schmidtke et al. 1996). The activation of the active sites is an autocatalytic event that takes place during the final steps of proteasome assembly, securing the cell against a premature activation of the enzyme complex (Schmidtke et al. 1996; Nandi et al. 1997; Krüger et al. 2001).

Since the three active sites within one β ring differ in their proteolytic properties, the 20S proteasome has been termed a multicatalytic protease (Wilk and Orlowski 1983). These differences have been shown using short fluorogenic peptide substrates. Accordingly, the activities have been named chymotrypsin-like, trypsin-like, postacidic, or caspase-like (Dahlmann et al. 1986; Cardozo et al. 1993; Orlowski et al. 1993; Kisselev et al. 1999). However, the cleavage preferences of the different active sites are far from absolute since cleavage of a given peptide bond within a longer substrate seems to depend on the quality of neighboring residues as well as on their interplay with residues which are more distant from the actual cleavage site (Shimbara et al. 1997; Nussbaum et al. 1998). Not neglecting certain preferences, cleavage site usage by the proteasome is promiscuous in that almost every amino acid residue can serve as a cleavage site giving the enzyme complex a high degree of flexibility with regard to quality of the products that are generated (Kloetzel 2001).

An important feature of proteasome-dependent protein breakdown is that as proteins thread through the different cavities their degradation is highly processive, resulting in the generation of polypeptides and peptides of 3–22 residues with a mean length of approximately 8–9 residues (Kisselev et al. 1998; Toes et al. 2001; Rock et al. 2002). Whether there is cooperation between the different active sites is still a matter of debate, although more recent detailed kinetic investigations indicate that protein processing is a noncooperative event, and it has been postulated that there exist so-called modifier sites that are involved in the ordered breakdown of protein substrates (Schmidtke et al. 2000; Kisselev et al. 2002).

The 26S proteasome

The 26S proteasomes is formed through the ATP-dependent association of two 19S complexes with the two outer α rings of the 20S core (Peters et al. 1993; Voges et al. 1999). The 19S complex, also called 19S regulator (Ferrell et al. 2000), is made up by approximately 18 subunits consisting of two functionally different entities (Glickman et al. 1998). The base that binds to the two outer α rings of the 20S core is formed by six different ATPases of the AAA family (ATPases associated with a variety of cellular activities) and two additional non-ATPase subunits. The base is responsible for the ATP-dependent opening of the central gate and hence the activation of the 20S core (Glickman et al. 1999) as well as the reverse-chaperone-like unfolding of protein substrates (Braun et al. 1999; Strickland et al. 2000). The upper part of the 19S regulator that is not in direct contact with the 20S core is called the lid (Fig. 1). The lid, which provides the specificity to proteolysis, is composed of at least ten non-ATPase subunits and contains the binding sites for ubiquitinated substrates, nonubiquitinated substrates, and also enzymes for the breakdown and recycling of ubiquitin chains. Although the base is sufficient for unfolding a protein substrate and

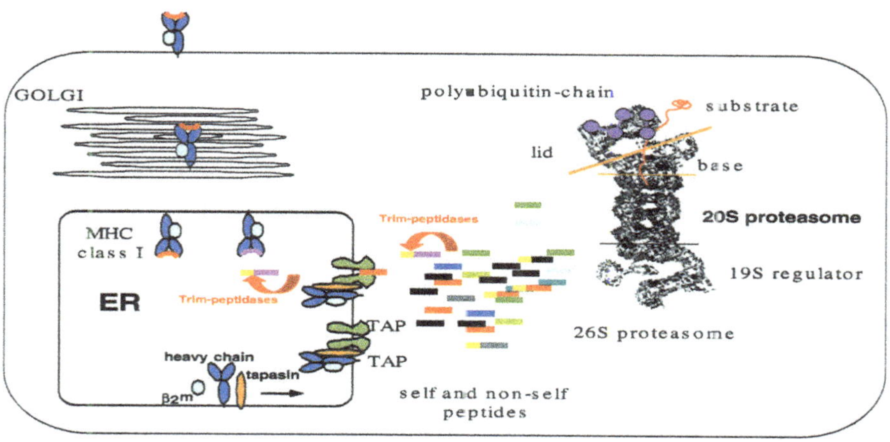

Fig. 1 The proteasome and antigen processing. Substrates that harbor MHC class I epitopes are tagged with a multiubiquitin chain. Tagged substrates are targeted to the 26S proteasome for degradation. Proteasomal processing results in the generation small peptides which are transported into the endoplasmatic reticulum (*ER*) by the transporters associated with antigen presentation (*TAP*). In case of N-terminal extensions these peptides may be trimmed either in the cytosol or in the ER by so-called trim peptidases. Within the ER, peptides bind and stabilize the appropriate MHC class I molecules. Fully assembled and loaded MHC class I molecules are translocated to the cell surface. By binding specifically to an MHC molecule encoded by a specific haplotype loaded with a unique peptide, cytotoxic T cells with complementary T-cell receptors (TCRs) are stimulated to proliferate and destroy the infected target cell

for opening the gate, the lid components are thought to play a role in the recognition of polyubiquitinated proteins and are essentially required for their degradation.

The proteasome regulator PA28

Biochemical screens have identified several additional regulatory molecules. The best characterized is the proteasome activator PA28/11S regulator, a heptameric 180–200 kDa complex, composed of α and β subunits (Dubiel et al. 1992; Ma et al. 1992). PA28 attaches in an ATP-independent way to the outer α rings of the 20S proteasome and in vitro strongly stimulates the hydrolysis of short fluorogenic peptide substrates. The turnover rate of cellular native proteins appears, however, not to be affected. Detailed kinetic analysis showed that PA28 activates the 20S proteasome without affecting the active sites by either facilitating substrate entry or product exit (Stohwasser et al. 2000b).

A structural explanation for the observed kinetic effects of PA28 binding was obtained by the crystal structure of a complex between the trypanosome 11S regulator PA26, and the yeast 20S proteasome (Knowlton et al. 1997; Whitby et al. 2000). It has to be kept in mind, however, that this is a noncognate complex formed between the 20S proteasome from yeast, which does not possess PA28- or PA28-like molecules, and that the trypanosome 11S regulator appears not to activate its cognate 20S proteasome (Yao et al. 1999). Nevertheless, there are good reasons to regard these structural data as being highly relevant for our principle understanding of PA28 function (Hill et al. 2002).

The PA26/20S core complex crystal structure reveals that PA26 induces conformational changes via its C-termini in the N-terminal parts of the α subunits with the effect that the central gates in the two outer α rings of the 20S core complex are opened. Based on this and on the observation that PA28 facilitates product exit, it was proposed that such an open conformation might allow the release of slightly longer peptides and in consequence might support the presentation of MHC class I antigen presentation.

The data obtained from X-ray structure analysis are supported by mutational and structural analysis of yeast proteasomes. A deletion of the N-terminal 9 amino acids of the $\alpha 3$ subunits or a mutation within the N-terminus of the $\alpha 3$ subunits results in the complete opening of the central gate controlled by the N-termini of the α subunits and the full activation of the otherwise latent or inactive proteasome (Groll et al. 2000; Köhler et al. 2001). Based on these studies, the so-called gating model finds widespread acceptance at the moment. This is due to the fact that it appears to be in agreement with most of the available in vitro data and with our understanding that the central gate has to be opened to allow access of the substrates to the catalytic cavity, notwithstanding also the release of the products from the proteasome.

One important and often overlooked aspect of the current gating model is that opening of the gate by regulatory molecules, i.e., 19S regulator and PA28, does not induce structural changes on the 20S proteasome, with the effect that the basic proteolytic activity within the β rings is changed. In contrast, the measured proteolytic activity using short fluorogenic peptide substrates correlates directly with the width of the gate (Stohwasser et al. 2000b; Kopp et al. 2001). Nevertheless, based on mutational analysis of PA28γ/Ki, a sequence homologue of PA28$\alpha\beta$, Rechsteiner and coworkers suggested that binding of the mutated PA28γ/Ki to the α rings results in a specific allosteric activation of certain active sites (Li et al. 2001). Although the authors do not explicitly dispute that gating affects proteasome activity, they inferred that the observed changes in proteasome activity are mainly due to regulator-induced allosteric changes in specific active sites. Recent data obtained from tissue culture experiments support Rechsteiner's conclusion that PA28 induces structural changes on the proteasome (Sijts et al. 2002; Sun et al. 2002). From these data it is, however, inferred that the binding of PA28$\alpha\beta$ in vivo and in vitro may effect substrate channelling and the accessibility of active sites for a given natural substrate rather than affecting the characteristics of the active sites themselves (see later sections).

Since proteasome-dependent cellular protein breakdown in general requires ATP and the presence of the 19S regulator, the importance of PA28/20S complexes in vivo is still unclear. Recent investigations showed that PA28 in connection with the 20S core and the 19S regulator may form so-called hybrid proteasomes (Hendil et al. 1998; Tanahashi et al. 2000; Kopp et al. 2001; Kisselev et al. 2002) Interestingly, these PA28-20S-19S hybrid proteasomes exhibit a proteolytic activity against fluorogenic peptide substrates which is almost identical to that of the 19S-20S-19S proteasome. Thus, one can conclude that at least one of the functions of the two different types of proteasome regulator is the opening of the central gates in the two outer α rings of the 20S core complex whereby both PA28 and the 19S regulator appear to open the gate to its maximal width.

The PI31 protein

The same biochemical screens that allowed the identification of PA28 also resulted in the identification of a molecule called PI31 that inhibits proteasome function. PI31 is a protein of approximately 31 kDa that presumably functions as homodimer. In vitro PI31 inhibits 20S-mediated cleavage of short fluorogenic substrates and of polypeptides and competes the binding of PA28 to 20S proteasomes (Zaiss et al. 1999; McCutchen-Maloney et al. 2000). This suggests that PI31 binds the α rings of the 20S proteasome and thereby obstructs the access to the catalytic cavity. Nevertheless, transfection of PI31 into intact cells recently showed that rather than inhibiting proteasome function PI31 acts as a selective modulator of the proteasome-mediated steps in MHC-class I antigen processing. Overexpression of PI31 had no impact on proteasome-mediated proteolysis but interfered selectively with the maturation of immuno-proteasome. The in vivo function of PI31 may therefore differ from the previously proposed function as a general inhibitor of proteasome activity (Zaiss et al. 2002).

Biogenesis of proteasome complexes

As pointed out above, proteasome-dependent proteolysis is in part regulated at the level of subunit incorporation and based on more recent results also at the level of gene expression. In eukaryotes, proteasome biogenesis is a precisely ordered multistep event involving the biosynthesis of all subunits, their assembly and maturation processes.

The first step is the biosynthesis of seven different α-type and the proforms of seven different β-type subunits of the 20S core complex. For a long time, this step has been neglected as a possible control mechanism for the amount of proteasomes in the cell although it is known that gene expression of ATP-dependent proteolysis systems is strongly regulated in bacteria (Gottesman 1996; Krüger et al. 1996; Missiakas et al. 1996). For instance, genes encoding Clp proteases in gram-positive bacteria are negatively controlled by a repressor which is degraded by the ClpCP protease under inducing conditions, resulting in a positive autoregulatory feed-back loop (Krüger and Hecker 1998; Krüger et al. 2001). A first indication for a coordinate control of eukaryotic proteasomal gene expression was reported for the baker yeast *Saccharomyces cerevisiae*. Using bioinformatics approaches for the whole yeast genome, a common upstream activating *cis*-element could be identified in front of genes encoding proteasomal subunits (Mannhaupt et al. 1999; Jensen and Knudsen 2000). This nonamer box was referred to as PACE element (proteasome associated control element) (Mannhaupt et al 1999). The Rpn4 protein formerly found as a ligand of the 26S proteasome (Glickman et al. 1998; Verma et al. 2000) was determined to be a transcription factor that activates proteasomal gene expression and recognizes PACE as a target sequence (Mannhaupt et al. 1999). Following that observation it was also shown that Rpn4 is essentially required for normal levels of proteasomal subunits and thereby for normal intracellular proteolysis. These results underline the function of Rpn4 as a general proteasomal transcription factor. Moreover, they demonstrated that Rpn4 is not only a transcriptional regulator for the 26S proteasome, but also an extremely short-lived protein. Once a certain level of newly formed proteasomes is available for the cell, Rpn4 becomes a substrate of the 26S proteasome. Such a regulatory mechanism yields in a negative feed-back loop in which the protein that induces proteasome biogenesis is destroyed by the fully assembled and matured proteasome in yeast (Xie and Varshavsky 2001). Recently, the global cellular response to proteasome inhibition was studied for the yeast system using whole-genome technologies. Transcriptional profiling revealed a concerted Rpn4-dependent upregulation of all proteasomal subunits upon treatment with proteasome inhibitors. These data do not only suggest Rpn4 as a master regulator responsible for the cell's ability to compensate for proteasome inhibition (Fleming et al. 2002), but also imply that the activity of the proteasome might be a sensor for the fine-tuning of the Rpn4 regulatory feed-back loop.

Although gene expression is also the first step in mammalian proteasome biogenesis and thereby its supposition, the knowledge available on its regulation is still very limited. Upon treatment of mammalian cells with proteasome inhibitors, the cells respond with a concerted upregulation of proteasomal subunits, resulting in the complete de novo onset of the biogenesis pathway. These data suggest the existence of an autoregulatory feed-back mechanism for compensation of inhibition of the proteasome activity also for mammalian cells (D. Heyken and E. Krüger, unpublished results).

Current models of early assembly events are based on the knowledge of simple proteasome structures like that of the archaebacterium *Thermoplasma acidophilum* that consists of only one α-type and one β-type subunit, or that of the eubacterium *Rhodococcus erythropolis* bearing two different α-type and two β-type subunits (Dahlmann et al. 1989; Zwickl et al. 1994; Zühl et al. 1997). The initial assembly step is thought to be characterized by the spontaneous formation of a heptameric α ring which then serves as template onto which the β subunit precursors are docked (Zwickl et al. 1994). For systems with more than one α and β subunit like *R. erythropolis* an alternative model suggests that the assembly of α/β heterodimers is necessary for the completion of the entire process (Mayr et al. 1998).

In eukaryotes, the proteasome assembly pathway appears to be a tightly regulated and much more complex process which must be able to orchestrate the correct positioning of two sets of seven different α and seven different β subunits. This is supported by in vitro

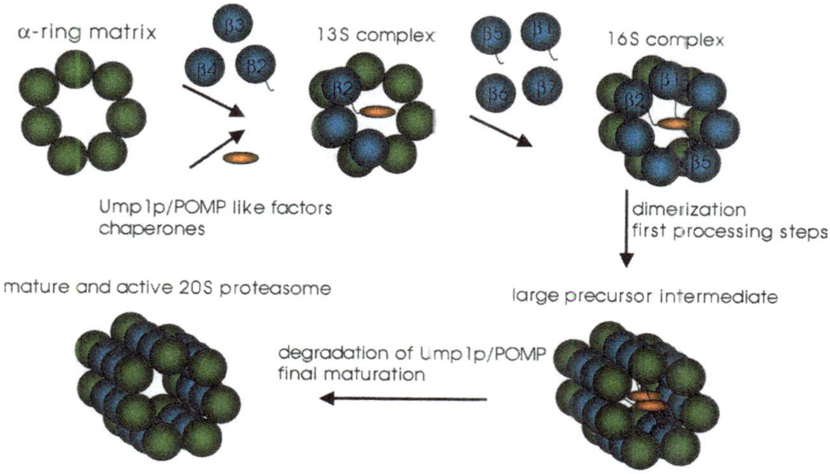

Fig. 2 The proteasome assembly pathway. Proteasome formation is initiated by the cooperative formation of the heptameric α ring, which serves as matrix for the docking of a defined set of β subunits. Structural rearrangements allow the binding of the residual subunits. The prosequences of the β subunits have specific functions during this process and are removed by *cis*- and *trans*-autocatalysis. This last step in proteasome maturation resulting in the liberation of the active sites probably occurs in the completely assembled cylinder

experiments with the human α7/C8 subunit which is able to form heptameric double ring structures if expressed alone or in combination with α1/Iota or α6/C2 whereby the two latter do not form ring structures on their own (Gerards et al. 1997; 1998). However, this α7/C8 subunit is not able to assemble into proteasome-like particles in the presence of the β subunits $\beta1/\delta$ or β7/N3 (Gerards et al. 1997). These authors therefore suggested a model of sequential $\alpha\beta$ interactions for the correct positioning of all subunits with α7/C8 having a central role in early assembly events.

In eukaryotic cells, the following assembly steps are determined by two distinct precursor intermediates resembling partially assembled complexes of all seven α subunits and some β subunits with sedimentation coefficients of approximately 13S and 16S. The β subunits β2/Z, β3/C10, and β4/C7 were identified as components of the 13S precursor. Later, the β subunits $\beta1/\delta$, β5/MB1, β6/C5, and β7/N3 are incorporated into the complex forming the 16S precursor intermediate, indicating that the subsequent association of β subunits occurs in several steps (Frentzel et al. 1994; Nandi et al. 1997; Schmidtke et al. 1997). At present, it is assumed that two 16S complexes interact to form a large and probably short-lived precursor intermediate, in which the final maturation steps take place (Fig. 2).

After cytokine exposure, the three active β subunits $\beta1/\delta$, β2/Z, and β5/MB1 can be replaced by their interferon (IFN)-γ-inducible counterparts β1i/LMP2, β2i/Mecl1, and β5i/LMP7 resulting in so-called immuno-proteasomes. This exchange requires the de novo assembly of the 20S core complex (Aki et al. 1994; Frentzel et al. 1994; Groettrup et al. 1997; Nandi et al. 1996; Griffin et al. 1998; Schmidt et al. 1999). The incorporation of the immuno-subunits is cooperative as β2i is only incorporated if β1i is present while β1i incorporation is independent of β2i (Groettrup et al. 1997). The β5i subunit seems to influence and support the kinetics of immuno-proteasome formation (Griffin et al. 1998). Due to this process, the proteolytic characteristics of the proteasome can be modulated representing an important physiological mechanism for the generation of antigenic peptides for the MHC class I-mediated immune response (Aki et al. 1994; Boes et al. 1994; Yang et al. 1995; Eleuteri et al. 1997; Groettrup et al. 1997; Kloetzel 2001 for review).

The complexity of proteasome assembly and the number of different subunits that have to be brought together for particle formation raises the question where within eukaryotic cells proteasome formation does take place. Localization studies revealed that mammalian 20S proteasomes are localized in the cytoplasm, at the ER as well as in the nucleus (Palmer et al. 1994; 1996; Peters et al. 1994; Wojcik et al. 1995; Knuehl et al. 1996; Reits et al. 1997). Furthermore, it has been shown for immuno-subunits that they almost exclusively localize to the ER (Brooks et al. 2000). These studies, however, did not distinguish between mature and immature proteasome particles and it remained unresolved whether the ER represents the site of proteasome assembly. For the yeast system, there is good evidence that at least for the biogenesis of nuclear 20S proteasomes specific assembly sites must exist at the nuclear envelope/ER network. In fact, nuclear proteasomes are imported as precursor complexes via the importin/cNLS receptor pathway (Enenkel et al. 1998; Lehmann et al. 2002).

With the exception of β3/C10 and β4/C7, all proteasomal β subunits are synthesized as inactive precursor proteins with N-terminal propeptides. These propeptides are processed by *cis*- or *trans*-autocatalysis in a two-step mechanism (Schmidtke et al. 1996). In fact, the propeptides of the active β subunits are cleaved off *cis*-autocatalytically, while the propeptides of the inactive subunits are removed by their active neighbors in *trans* (Schmidtke et al 1996; Chen and Hochstrasser 1996; Heinemeyer et al. 1997; Jäger et al. 1999; Salzmann et al. 1999). This final step in proteasome maturation has been suggested to occur during the late stages of assembly within the combined 16S complexes which were shown to be processing-competent (Chen and Hochstrasser, 1996; Heinemeyer et al. 1997; Schmidtke et al 1997).

For *Thermoplasma*, the N-terminal propeptides of β subunits are dispensable for proteasome biogenesis (Zwickl et al. 1994). From the *Rhodococcus* model system there are several lines of evidence that define the role of the propeptides in supporting the initial folding of the β subunits and promoting the maturation steps (Zühl et al. 1997). In eukaryotes, our studies and those of other laboratories show that without being absolutely essential for mammalian proteasome assembly the N-terminal propeptides influence the efficiency and timeliness of subunit incorporation and maturation to different degrees. The absence or exchange of the β5 prosequence exerts the strongest effect, which reflects a hierarchy of active site function as follows: $\beta5 >> \beta2 > \beta1$ (Chen and Hochstrasser 1996; Heinemeyer et al. 1997; Schmidtke et al. 1997; Jäger et al. 1999; Kingsbury et al. 2000). In yeast, the presence of the β5 prosequence is essential and was shown to possess an intramolecular chaperone-like function (Chen and Hochstrasser 1996). Therefore, incorporation of β5i (LMP7) into precursors and to a low extent into mature proteasomes in the absence of the prosequence may suggest that the prosequences are not essential for subunit targeting in mammalian cells. Furthermore, its function can to a large extent be taken over by the prosequence of the LMP2 subunit (Witt et al. 2000). Nevertheless, the β5i (LMP7) prosequence which can also function *in trans* was shown to mediate preferential incorporation of LMP7 into immuno-proteasomes but not into constitutive proteasomes. Thus, the prosequence clearly supports correct maturation and efficient incorporation into the right complex (Kingsbury et al. 2000; Witt et al. 2000).

Importantly, proteasome biogenesis in eukaryotes is assisted by accessory proteins that appear to promote the final assembly and maturation steps. For example, chaperones of the Hsp70 class bind to 16S proteasome precursor complexes and may keep proteasome assembly intermediates in a partially unfolded state to allow subunit processing, correct folding, and incorporation of late proteasome subunits (Schmidtke et al. 1997). Beside these chaperones, other accessory factors, namely the yeast maturation protein Ump1p (Ramos et al. 1998) and its mammalian homologues referred to as h/mUmp1, proteassemblin, or proteasome maturation protein POMP (Burri et al. 2000; Griffin et al. 2000; Witt et al. 2000), were also described to be associated to proteasome precursor complexes. These factors form part of the proteasome precursor complexes but not of mature 20S proteasomes, and become degraded upon final proteasome maturation (Frentzel et al. 1994; Ramos et al. 1998; Burri et al. 2000; Griffin et al. 2000; Witt et al. 2000). Deletion of the yeast Ump1p does not result in cell death as known for deletion of any proteasomal gene, but causes a severe stress-sensitive phenotype and the accumulation of precursor complexes. Mostly based on genetic experiments, the coordination of the β subunit processing, in particular by a physical interaction of Ump1p with the $\beta5$ prosequence, is discussed as a general physiological function for Ump1/POMP-like proteins (Ramos et al. 1998). Although the biogenesis process triggered by Ump/POMP-like factors seems to represent a conserved and general pathway in eukaryotic cells, the molecular mechanisms of this orchestration are not well understood. Probably due to the low similarity between yeast Ump1p and its mammalian homologues, these proteins are not functionally interchangeable (Burri et al. 2000).

Interestingly, gene expression of the yeast *UMP1* gene is regulated in a Rpn4-dependent manner at the level of transcription as it described for all genes encoding proteasomal subunits in yeast (Jensen and Knudsen 2000; Fleming et al. 2002). This supports the idea that balanced Ump1p levels are necessary for balanced proteasome levels and thereby for balanced intracellular protein homeostasis. In this context, it is of interest that gene expression of mammalian Ump1/POMP-like factors appears to be differentially regulated as well. The mRNA level of the human and mouse homologues is induced after treatment with IFN-γ, suggesting that these factors are somehow involved in immuno-proteasome biogenesis (Burri et al. 2000; Griffin et al. 2000; Witt et al 2000).

The proteasome system is a component of the immune system

Every organism is constantly confronted with numerous different pathogens like viruses, bacteria, and parasites. In higher vertebrates, most of the pathogens are attacked immediately after their penetration by an antigen-unspecific mechanism of the innate immunity system. In case a pathogen cannot be completely eliminated within a short time span, the adaptive immune response takes over. This antigen-specific immune reaction results in clonal expansion and differentiation of T and B cells. B cells secrete antibodies for recognition of pathogens in the extra cellular space. T-cells detect intracellular pathogens and participate in the activation of other cells of the immune system.

The fundamental difference in the function of $CD4^+$ and $CD8^+$ T cells is reflected by the way the antigens which they recognize are processed. Microbial antigens that are taken up by professional antigen-presenting cells via endocytosis or phagocytosis are primarily processed within acidified endosomes. There the proteolytically processed antigens bind to MHC class II molecules. In contrast, proteins of viruses or other pathogens that are localized in the cytosol of the infected cells are processed in the cytosol by the proteasome system. Peptides of appropriate length and amino acid composition bind the transporter associated with antigen processing (TAP) for translocation into the endoplasmic reticulum (ER) and then to MHC class I molecules.

The properties of MHC class I antigens

The generation of peptide-loaded MHC class I molecules mostly requires the proteolytic generation of peptides with a preferred length of 8–10 amino acids in the cytosol, the efficient transport of peptides via TAP-proteins (transporter associated with antigen processing) into the endoplasmic reticulum (Fig. 1). There, peptide-induced stabilization and the assembly of MHC class I trimers takes place followed by transport of the peptide-loaded MHC molecules to the cell surface. Peptide binding to the MHC class I protein occurs through the peptide's side chains with pockets in the peptide binding groove of the MHC protein (Madden 1995). The specificity of this interaction is mediated by so-called anchor residues. One of these anchor residues is always located at the carboxy-terminus of the peptide, fixing the peptide to the edge of the MHC-peptide binding groove. The other one or two anchor residues are usually located near the N-terminus of the antigenic peptide. To allow peptide binding, the anchor residue positions must be occupied by specific residues defined by the haplotype of the MHC class I molecules. The carboxy-terminal anchor residues are either hydrophobic (L, I, Y, V, or M) or basic (K or R), while at the second anchor also acidic residues (D or E) or aromatic residues (N, Y, F) can be found. In addition, several of the other residues can make contact with the peptide binding groove and influence the stability of the peptide interaction. Thus, within a given MHC sequence motif certain residues are either preferred or disfavored (Falk et al. 1991). The fact that the carboxy-terminal anchor residues are enriched for hydrophobic or basic residues suggests the existence of a proteolytic machinery with a certain predisposition for generating such peptides. Based on genetic, molecular genetic, and biochemical experiments, it is now accepted that the proteasome system is responsible for the generation of the majority of MHC class I ligands (Kloetzel 2001; Rock et al. 2002). In specific cases, other proteases such as TPPII or TOP might also contribute to the pool of antigenic peptides (Goldberg et al. 2002; Saveanu et al. 2002; Levy et al. 2002). However, due to their selective cleavage specificity these proteases will only generate a limited subset of MHC class I peptides.

IFN-γ-inducible components of the proteasome system

The first indication that proteasomes may play an important role in the generation of MHC class I-presented peptides came from the finding that two proteasome subunits are encoded in the MHC class II region in direct vicinity to the TAP-genes and that their expression is inducible by IFN-γ(Glynne et al. 1991; Brown et al. 1991; Kelly et al. 1991; Ortiz-Navarette et al. 1991; Martinez and Monaco 1991; Yang et al. 1995). Stimulation of cells with IFN-γ also induces the synthesis of the above-described α and β subunits of the proteasome activator PA28 (Nandi et al. 1997). Interestingly, the expression profile of the two subunits largely correlates with that of the immuno-subunits (Macagno et al. 1999). In concert with the 19S regulator, these five IFN-γ-inducible components of the proteasome system form the so-called immuno-proteasome, indicating its close connection with the cellular immune response (Aki et al. 1994; Hendil et al. 1998).

The role of proteasome immuno-subunits in MHC class I antigen processing

As described above, under conditions of IFN-γ induction the entire catalytic machinery of the proteasome complex is replaced. In fact, cells with antigen-presenting functions as found in the thymus, spleen, and lymph nodes express the IFN-γ-inducible subunits constitutively (Macagno et al. 1999; Stohwasser et al. 1997; Eleuteri et al. 1997). Furthermore, also other stimuli like TNF-α and IFN-β as well as stimulation of dendritic cells can influence the expression levels of the three immuno-subunits (Jamaluddin et al. 2001; Kuckelkorn et al. 2002a).

Although the inducibility by IFN-γ strongly suggested that the three immuno-subunits must play an important role in MHC class I antigen processing, their exact function in this process has long been questioned (Arnold et al. 1992; Momburg et al. 1992) and it has still not been fully resolved. First experimental attempts to demonstrate a connection between the proteasome system and MHC class I antigen processing showed that purified constitutive as well as immuno-proteasomes are able generate a correct 9mer epitope derived from the IE pp89 protein of the murine cytomegalovirus (Boes et al. 1994). In addition, Michalek et al. (1993) had presented evidence that defects in the ubiquitinylation pathway disturbed antigen processing of chicken albumin. The availability of membrane-permeable proteasome inhibitors furthermore allowed demonstration of a direct correlation between cellular proteasome activity and the production of antigenic peptides that bind to the antigen binding groove of MHC class I molecules (Rock et al. 1994; Harding 1996). Demonstrating a direct correlation between proteasome activity and the generation of MHC class I-presented peptides, it was shown that in cells treated with proteasome inhibitors, the majority of MHC class I molecules was retained in the ER due to shortage of peptide supply (Rock et al. 1994).

In vitro digestion of polypeptide substrates encompassing antigenic peptides by purified 20S proteasomes, and studies of antigen presentation with intact cells which were transfected with N-terminally and C-terminally extended antigenic peptides showed that proteasomes liberate the correct C-terminus of an antigenic peptide (Boes et al. 1994; Dick et al. 1996; Eggers et al. 1995; Niedermann et al. 1997). Cleavage at the N-terminus of an epitope is, however, less precise. In many cases, N-terminally extended epitope precursor peptides are generated that are trimmed to the correct size by either ER-resident or cytosolic trim-peptidases (Ossendorp et al. 1996; Craiu et al. 1997; Beninga et al. 1998; Stoltze et al. 2000; Knuehl et al. 2001; Serwold et al. 2002; Saric et al. 2002).

To elucidate the role of IFN-γ-inducible subunits in the generation of antigenic peptides, constitutive and immuno-proteasomes were purified from cells and tested with respect to their activity against different fluorogenic peptide substrates (Driscoll et al. 1993; Gaczynska et al. 1993; Boes et al. 1994; Kuckelkorn et al. 1995; Ustrell et al. 1995; Eleuteri et al. 1997). All these studies showed a clear reduction of the caspase-like activity upon incorporation of the immuno-subunits and due to the replacement of $\beta 1$ by $\beta 1i$. However, inconsistent changes in the activities displayed by the other two β subunits, i.e.,

Table 1 Effects of immunoproteasomes and PA 28 on MHC class 1 antigen processing

Antigen	Epitope	Effect of immuno-subunits	Effect of PA 28
Ovalbumin	257–264	No effect	n.d.
HY antigen	n.d.	+	n.d.
Influenza A nucleoprotein	366–374	+	n.d.
Murine cytomegalovirus pp89	168–176	n.d.	+
JAK 1 tyrosine kinase	355–363	n.d.	+
Influenza A/PR/8 nucleoprotein	146–154	n.d.	+
Hepatitis B virus core antigen	141–151	+	n.d.
Adenovirus E1A	234–243	No effect	No effect
Adenovirus E1B	192–200	+	n.d.
LCMV nucleoprotein	118–126	+	No effect
Moloney murine leukemia virus gag pr 75	75–83	+	+
Moloney murine leukemia virus env gp70	189–196	No effect	No effect
RU1 tumor antigen	34–42	–	n.d.
Melan-A tumor antigen	26–35	–	n.d.
Tyrosinase related protein 2 TRP-2 tumor antigen	360–368	No effect	+
Tyrosinase related protein 2 TRP2 tumor antigen	288–296	–	n.d.

$\beta5/\beta5$i-chymotrypsin-like, and $\beta2/\beta2$i trypsin-like, were observed. Furthermore, detailed analysis of cleavage products generated from larger polypeptides and protein substrates by both constitutive and immuno-proteasomes demonstrated that activities of the three different catalytic centers towards longer substrates are less well defined (Nussbaum et al. 1998; Schmidtke et al. 1997). In other words, while cleavage after basic residues in fluorogenic peptide substrates (lysine/arginine) are exclusively mediated by the $\beta2/\beta2$i pair of catalytic subunits (Salzmann et al. 1999), cleavages after the same residues within polypeptide substrates appear to be performed by different active site subunits. In addition, the analysis of mutant cell lines with defective LMP2 and LMP7 expression or expression of mutant, inactive LMP2 did not reveal any consequences for antigen presentation (Momburg et al. 1992; Arnold et al. 1993; Yewdell et al. 1994; Schmidtke et al. 1998). On the other hand, it was found that LMP7 expression was essential for the processing of an influenza virus-derived CTL-epitope (Cerundolo et al. 1995).

More detailed investigations over the recent years have changed the initial skeptical view on the functional importance of immuno-subunits. Transfection experiments with cDNA encoding $\beta1$i, $\beta2$i and $\beta5$i in connection with extensive analyses of CTL epitope generation demonstrated that the presence of immuno-subunits enhances the presentation of a major subset of virus-derived antigen peptides (Schwarz et al. 2000b; Sijts et al. 2000a; Sijts et al. 2000b; van Hall et al. 2000), although this effect does not apply to all viral epitopes (Table 1).

Despite this accumulating information on the role of the immuno-subunits in MHC class I antigen presentation, the mechanisms that are involved are not fully understood. In vitro digestion experiments with purified 20S proteasomes and synthetic polypeptides encompassing the antigenic peptides and natural flanking sequences as substrate confirmed in almost all cases the observations made in intact cellular systems (Sijts et al. 2000a; Schwarz et al. 2000a, 2000b). Analysis of the in vitro-generated peptide fragments showed that the presence of immuno-subunits altered the cleavage site preference of proteasomes. This resulted in a more frequent usage of specific cleavage sites, with the consequence that the relative abundance of certain peptides within the generated peptide pool is changed.

Whether or to which degree, the presence of immuno-subunits also changes the quality of cleavage sites used is at the moment difficult to access.

Interestingly, experiments in which Hela cells were infected with vaccinia virus expressing the hepatitis B virus (HBV) core antigen showed that the HBVcAg $_{141-151}$ was only efficiently liberated after IFN-γ stimulation. Furthermore, although the concerted presence of all three immuno-subunits seemed to be required, an inactive β5i subunit, in which the Thr1 was mutated to Ala rendering this subunit inactive, also supported epitope generation. These experiments demonstrated that the altered cleavage profile found to be associated with immunosubunit incorporation was due not only to cleavage specificity, but most likely also to structural changes imposed on the proteasome complex by the incorporation of the immuno-subunits (Sijts et al. 2000a). The idea that the primary effect of immuno-subunit incorporation is the alteration of proteasome structure is also supported by the changed chromatographic properties of immuno-proteasomes that is caused by changes in their surface charge (Dahlmann et al. 2000).

Interestingly, a negative effect of immunosubunit expression on viral CTL epitope production was observed in none of the analyses referred to above. In contrast, recent studies have indicated that this may be different in case of self antigen-derived CTL-epitopes (Table 1). Morel et al. (2000) reported that the incorporation of immuno-subunits into the proteasome abrogated the production of two antigenic peptides, one derived from the self protein RU1 and one derived from the melanoma differentiation antigen Melan-A. On the other hand, we recently observed that such a down regulation of tumor epitope generation upon exposure to IFN-γ can also occur in the absence of induction of immuno-subunit expression (Sun et al. 2002; Sijts et al. 2002). Thus, at least in this case, the downregulation of epitope presentation must be caused by IFN-γ-induced factors other than the proteasome.

The role of PA28 in MHC class I antigen processing

Stimulation of cells with IFN-γ not only induces the expression of immuno-subunits, but also that of PA28$\alpha\beta$. Although PA28, unlike the immuno-subunits, is found in almost any cell type, even in the absence of cytokine stimulation, enhanced constitutive levels are observed in cells with specialized antigen-presenting function (Macagno et al. 1999). Cell systems that express both PA28 subunits independently of IFN-γ were used to study the role of PA28 in antigen processing. The investigation of viral antigens showed that PA28 enhances the presentation of some viral epitopes without increasing overall protein turnover or the turnover of viral protein substrates (Groettrup et al. 1996; van Hall et al. 2000), while the presentation of other virus-derived epitopes was not effected (Table 1). Furthermore, this enhanced peptide presentation is independent of the presence of immuno-subunits in the 20S proteasome (Stohwasser et al. 2000a; Schwarz et al. 2000a), excluding the possibility that PA28 might exert its function by increasing immuno-proteasome formation as was proposed recently (Preckel et al. 1999). In vitro studies of proteasome-mediated liberation of antigenic peptides from polypeptide substrates in the presence of PA28 revealed an immediate liberation of PA28-dependent epitopes (Dick et al. 1996; Kuckelkorn et al. 2002b). Thus, similar to the immuno-proteasomes, PA28 does not affect presentation of all MHC class I epitope equally. Somewhat in contrast, PA28 seems to affect the generation and presentation of only a selective and minor fraction of epitopes.

Analysis of the effects of PA28 on proteasomal cleavage site usage reveals that PA28 does not induce new cleavage specificities. Furthermore, there is no evidence that PA28 has a major impact on substrate turnover rates. Instead, PA28 enhances the frequency at which specific cleavage sites are used, resulting in the immediate liberation of the peptides residing in between two of these preferred cleavage sites (Dick et al. 1996; Kuckelkorn et al. 2002; Sun et al. 2002).

The most striking example to date for PA28 dependency of an epitope was recently obtained when the processing and presentation of two CTL epitopes derived from a melanoma differentiation antigen, i.e., tyrosinase related protein 2 (TRP2) was analyzed. These studies showed that several melanoma cell lines failed to present one of the two antigenic peptides, i.e., TRP2 $_{360–368}$ and TRP2 $_{288–296}$. Detailed biochemical and molecular experiments showed that the inability of melanoma 18a cells to present the TRP2 $_{360–368}$ epitope correlated with a strongly impaired expression of PA28 in these cells, which also was not restored by IFN-γ treatment. Epitope presentation and the ability to activate TRP2$_{360–368}$ epitope-specific CTLs by melanoma 18a cells were, however, fully restored by transient transfection with cDNAs encoding the PA28α and PA28β subunits. Apart from demonstrating that PA28 plays an important role in the cellular immune response, these experiments demonstrated for the first time that PA28 expression is able to alter the immunological phenotype of a cell. In vitro experiments and biochemical analysis also demonstrated that only in the presence of PA28 20S were proteasomes able to liberate detectable amounts of the TRP2 $_{360–368}$ epitope, i.e., its N-terminally extended precursor peptides. Despite the drastic effects observed in vitro as well as in vivo, PA28 apparently did not influence the quality of cleavage site usage.

PA28 hypotheses

How then does PA28 exert its effect on antigen processing and presentation? Based on the structural data and the observation that PA28 facilitates product exit, it was proposed that an open conformation of the proteasome might allow the release of slightly longer peptides (Whitby et al. 2000). In this scenario, these peptides would be rescued from further degradation. This in turn may support the generation of certain CTL epitopes and in particular those that are produced from N-terminally extended precursor peptides.

In fact, it has been shown that even epitopes which are generated by the proteasome with correct size for MHC class I binding are often also produced with short N-terminal extension requiring trimming by aminopeptidases such as TOP or bleomycin hydrolase (Stoltze et al. 2000; Rock et al. 2002). Consequently, if PA28 indeed supports the generation of longer peptide, the result would not only be the generation of a new quality of peptides, but also the amount of peptide available would be increased.

However, since both the 19S regulator complex as well as PA28 induce the complete opening of the gate, it is obvious that this gating hypothesis, as attractive as it appears, does not fully explain the effects on substrate cleavage alone. Furthermore, if an increase in peptide length as a result of PA28 action would be of general benefit for MHC class I-dependent antigen presentation one would expect to find an increased MHC class I surface expression upon PA28 induction. However, such an effect has not been observed in any of the studies performed to date. Therefore, an alternative hypothesis which does not in principal exclude the above-discussed gating model would be that PA28 by binding to the outer α ring not only opens the gate at the site opposite to the 19S regulator binding sites, but also induces subtle structural changes which allow better access to the actives sites of those epitope-containing domains (for example, by gaining higher affinity to the substrate binding grooves) which otherwise possess only a strongly reduced affinity. Such a view would be in concordance with the observed selectivity of PA28 and data showing that PA28 effects the proteasomal usage of those cleavages which in its absence are less efficiently recognized (Sijts et al. 2002).

The proteasome system and tumor epitope generation

Under noninflammatory conditions, many cell types lack expression of immuno-subunits and express PA28 differentially and only at low levels (Soza et al. 1997). This suggests that a continuous presence of these molecules in all tissues is either not required or desired. In fact, neither the incorporation of immuno-subunits nor the upregulated expression of PA28 in transfectant cells seems to influence the antigen processing-independent functions of the proteasome system. Therefore, one possible function of immuno-subunits and PA28 may lie in the development of the $CD8^+$ cell repertoire. In particular, the observation that in some cases immuno-subunits may influence antigen processing of foreign and self-antigens differently appears indicative for this assumption (Morel et al. 2000). The expression of immuno-subunits in dendritic cells within the thymus may result in a negative selection of T cells reactive against peptides derived from self antigens, leaving only those T cells that react against peptides generated by constitutive proteasomes as potential targets for autoimmune CTL.

In this context, it is remarkable that Morel et al. (2000) recently reported that immuno-proteasomes are unable to produce several self antigen-derived CTL epitopes, including an important CTL epitope derived from Melan-A. These data have been the first example of CTL epitopes that are only generated by constitutive proteasomes and appear to be destroyed by immuno-proteasomes. At the same time, it must also be stated that even stimulated dendritic cells always contain a mixed population of proteasomes comprising constitutive as well as immuno-proteasomes (Macagno et al. 1999). Interestingly, Sun et al. (2002) recently also reported on a TRP2-derived tumor epitope whose surface presentation was downregulated in response to IFN-γ in Melan 18a cells. However, in that case it could be shown that this was not due to the presence of immuno-proteasomes since these cells lacked immuno-proteasomes and also did not respond with immuno-proteasome upregulation upon IFN-γ stimulation.

Synergistic effects

The observation that both PA28 as well as the immuno-subunits are induced by IFN-γ may suggest that they act synergistically within the cellular immune response. Is there any evidence for that? The evaluation of the effects of PA28 and immuno-subunits on epitope generation raises several interesting points. For example, the two proteins exhibit different effects on the processing of separate epitopes derived from one pathogen. Thus, while both enhance the presentation of a Moloney leukemia virus gag protein epitope, they do not effect the presentation of an epitope of the Mulv env protein (van Hall et al. 2000). In addition, it was shown that PA28 and immuno-proteasome can exert their effects independently. Thus, induction of immuno-proteasome formation in mouse embryonal cells results in an increased MHC class I expression of the Moloney leukemia virus gag protein epitope, in the absence of PA28 expression. Conversely, induction of PA28 synthesis improved the presentation of the gag epitope in the absence of immuno-proteasomes. A recent report on the analysis of PA28β-deficient mice suggested that PA28 may influence immuno-proteasome formation and thus increase the immuno-proteasome content of a cell (Preckel et al. 1999). On the other hand, such an effect was not observed in cell lines (Schwarz et al. 2000) and also not in PA28$\alpha\beta$ double knock-out mice (Murata et al. 2001).

The proteasome system: a target of immune escape

As the proteasome is the central generating source for antigenic peptides, it is not surprising that viruses and also tumor cells have evolved strategies to interfere with proteasome function. In a number of tumor types such as colon carcinoma, small cell lung carcinoma, pancreatic carcinoma, hepatocellular carcinoma, renal cell carcinoma, and melanoma, a downregulation of immuno-subunit expression was observed (Restifo et al. 1993; Seliger et al. 1996; Johnson et al. 1998; Delp et al. 2000). Although the absence of PA28 α and β seems to be less frequently effected, their complete absence has been reported for several melanoma cell lines (Seliger et al. 2001; Sun et al. 2002). In some cases, the expression of immuno-subunits and/or PA28$\alpha\beta$ expression appeared coordinately downregulated, while in other instances only single subunits were affected. In most cases, treatment with IFN-γ is restores the expression of the proteasome components but there are also exceptions (Sun et al. 2002). Altogether it appears that the suppression of PA28 and immuno-subunits in different tumor cells is effected by quite different regulatory mechanisms, and that their absence is advantageous for tumor cells growth by evading a CTL-mediated immune surveillance.

Direct functional evidence for viral interference with the proteasome system is only just beginning to emerge. In this context, it is important to realize that many viruses are dependent on proteasome activity for their survival. For instance, proteasomes process the HIV gag polyprotein into different gag proteins and that interference with this process impairs maturation of HIV particles (Schubert et al. 2000). Human papilloma virus (HPV) interferes with the proteasome system by various mechanisms. For example, the HPV16 E7 protein binds one of the six ATPase subunits of the base, thereby enhancing proteasome activity and inducing increased turnover of the Rb retinoblastoma tumor suppressor protein (Boyer et al. 1996; Berezutskaya and Bagchi 1997). In addition, HPV16 E6 targets the p53 tumor suppressor protein to proteasome-dependent degradation by recruiting the ubiquitin ligase E6-AP (Scheffner et al. 1993), thus contributing to oncogenic transformation of HPV infected cells.

Evidence for direct interaction with proteasome subunits has been obtained for the human T-cell leukemia virus typ1 (HTLV-1) Tax protein (Rousset et al. 1996), HBV pX protein (Fischer et al. 1995; Huang et al. 1996; Stohwasser et al. 2003) HIV Nef (Turnell et al. 2000), and Tat (Seeger et al. 1997), and adenovirus (Ad5) E1A. For the Tax protein, it was shown that it binds the N3(β7) and C9α3 subunits and also activates the transcription factor NF-kB. Both the HIV Tat and HBV pX protein bind to subunits of the α ring (α4) and interfere with the binding of PA28. Interestingly, recently it was shown that the binding of HIV Tat to the α rings also effects the capacity of PA28 to enhance antigen presentation (Huang et al. 2002).

In addition, both the Adenovirus type 12 E1A and HPV type 18 E7 proteins repress the β1i and/or β5i promoter and consequently subunit expression (Rotem-Yehudar et al. 1996; Berezutskaya and Bagchi 1997; Proffitt et al. 1997; Georgopoulos et al. 2000). A completely different strategy for proteasome inhibition is introduced by the Epstein Bar Virus nuclear antigen-1 (EBNA1). This protein contains an internal glycine/alanine repeat that hinders the proteasome-dependent degradation of this protein and the generation of EBNA1-derived CTL epitopes (Levitskaya et al. 1995; 1997).

Making use of the proteasome system

With regard to the development of epitope-based vaccines, a detailed knowledge of proteasomal cleavage properties is of great practical relevance as it would allow the identification of CTL epitope candidates within protein stretches. Kessler et al. (2001) first exploited such a "reverse immunology" method to characterize HLA-A2-presented CTL epitopes in PRAME a tumor-associated antigen that is expressed in a wide variety of tumors. The PRAME protein sequence was first screened for the presence of nona- and decamer peptides with HLA-A*0201 binding motif. Identified peptides were synthesized and tested for actual HLA binding. Putative epitope candidates found to bind with high affinity were synthesized as part of larger polypeptides, encompassing the potential antigenic sequences and their natural flanking residues, and offered substrates to purified 20S immuno-proteasomes. The analysis of the digestion products led to the identification of four epitope candidates that were generated by the proteasome. Each of the PRAME-derived peptides induced specific CTL responses in vitro and CTL clones established by this method specifically lysed PRAME-expressing HLA+0201 positive tumor cells.

To simplify CTL epitope identification strategies, computer prediction programs that predict cleavage site usage within proteins have been established. So far, two prediction programs based on different parameters have been published (Holzhütter and Kloetzel 2000; Kuttler et al. 2001). One program was trained using 20S cleavage data of the enolase-1 protein; the other program is based on polypeptide cleavage data generated in a large number of studies. Both programs predict proteasomal cleavages with high fidelity whereby the latter program not only identifies proteasome cleavage sites but also predicts the probability with which specific fragments are generated. Indeed, the application of this program in combination with a program that predicts MHC class I binding affinity led to the identification of eight CTL epitopes derived from the entire deduced proteome of *Chlamydia trachomatis* that are presented by HLA-B27 molecules in patients suffering from rheumatoid arthritis (Kuon et al. 2001).

In summary, these computer-assisted approaches enable a rapid and easy identification of CTL epitopes that can be included in the aimed CTL-inducing vaccine and will help the selection of epitope flanking spacers that improve proteasome-mediated epitope processing. The existing proteasome cleavage programs are currently tested for general applicability and the availability of new cleavage data obtained in vitro digestion of designed constructs with purified proteasomes in turn will allow the designers to improve their programs.

References

Aki M, Shimbara N, Takashina M, Akiyama K, Kagawa S, Tamura T, Tanahashi N, Yoshimura T, Tanaka K, Ichihara A (1994) Interferon-γ induces different subunit organizations and functional diversity of proteasomes. J Biochem 115:257–269

Arnold D, Driscoll J, Androlewicz M, Hughes E, Cresswell P, Spiess T (1992) Proteasome subunits encoded in the MHC are not generally required for the processing of peptides bound by MHC class I molecules. Nature 360:171–174

Beninga J, Rock KL, Goldberg AL (1998) Interferon-γ can stimulate postproteasomal trimming of the N-terminus of an antigenic peptide by inducing leucine aminopeptidase. J Biol Chem 273:187343–18739

Berezutskaya E and Bagchi S (1997) The human papillomavirus E7 oncoprotein functionally interacts with the S4 subunit of the 26 S proteasome. J Biol Chem 272:30135–30140

Boes B, Hengel H, Ruppert T, Multhaup G, Koszinowski U, Kloetzel PM (1994). IFN-γ stimulation modulates the proteolytic activity and cleavage site preference of 20S mouse proteasomes. J Exp Med 179:901–909

Boyer SN, Wazer DE, Band V. (1996) E7 protein of human papilloma virus-16 induces degradation of retinoblastoma protein through the ubiquitin-proteasome pathway. Cancer Res 56:4620–4624

Branninigan JA, Dodson G, Duggleby HJ, Moody PC, Smith JL, Tomchick, DR, Murzin AG (1995) A protein catalytic framework with an N-terminal nucleophile is capable of self-activation. Nature 378:414–419

Braun B, Glickman M, Kraft R, Dahlmann B, Kloetzel PM, Finley D and Schmidt M (1999). The base of the proteasome regulatory particle exhibits ATP-dependent chaperone-like activity. Nature Cell Biol 1:221–226

Brooks P, Murray RZ, Mason GG, Hendil KB, Rivett AJ (2000) Association of immunoproteasomes with the endoplasmic reticulum. Biochem J 352:611–615

Brown MG, Driscoll J, Monaco JJ (1991) Structural and serological similarity of MHC linked LMP and proteasome (multicatalytic proteinase complex). Nature 353:355–358

Burri L, Höckendorff J, Boehm U, Klamp T, Dohmen JR, Lévy F (2000) Identification and characterization of a mammalian protein interacting with 20S proteasome precursors. Proc Natl Acad Sci USA 97 10348–10353

Cardozo C (1993) Catalytic components of the bovine pituitary multicatalytic proteinase complex (proteasome). Enzymes Protein 47:296–305

Cerundolo, V, Kelly A., Elliott T, Trowsdale J, Townsend A (1995) Genes encoded in the major histocompatibility complex affecting the generation of peptides for TAP transport. Eur. J. Immunol 25:554–562

Chen P, Hochstrasser M (1996) Autocatalytic subunit processing couples active site formation in the 20S proteasome to completion of assembly. Cell 86:961–972

Craiu A, Akopian T, Goldberg A, Rock KL (1997) Two distinct proteolytic processes in the generation of a major histocompatibility complex class I-presented peptide. Proc Natl Acad Sci U S A 94:10850–10855

Dahlmann B, Kopp F, Kuehn L, Reinauer H, Schwenen M (1986) Studies on the multicatalytic proteinase from rat skeletal muscle. Biomed Biochim Acta 45:1493–1501

Dahlmann B, Kopp F, Kuehn L, Niedel B, Pfeifer G, Hegerl R, Baumeister W (1989) The multicatalytic proteinase (prosome) is ubiquitous from eukaryotes to archaebacteria. FEBS Lett 251:125–131

Dahlmann B, Ruppert T, Kuehn L, Merforth S, Kloetzel PM (2000) Different proteasome subtypes in a single tissue exhibit different enzymatic properties. J Mol Biol 303:643–653

Delp K, Momburg F, Hilmes C, Huber C, Seliger B (2000) Functional deficiencies of components of the MHC class I antigen pathway in human tumors of epithelial origin. Bone Marrow Transplant 25 Suppl 2:S88–95

Dick T, Ruppert T, Groettrup M, Kloetzel PM, Kuehn L, Koszinowski U, Stevanovic S, Schild HJ, Rammensee HG (1996). Coordinated dual cleavages induced by the proteasome regulator PA28 lead to dominant MHC ligands. Cell 86:253–262

Driscoll J, Brown MG, Finley D and Monaco JJ (1993). MHC-linked *LMP* gene products specifically after peptidase activities of the proteasome. Nature 365:262–264

Dubiel W, Pratt G, Ferrell K, Rechsteiner M (1992) Purification of an11S regulator of the multicatalytic proteinase. J Biol Chem 267:22369–22377

Eggers M, Ruppert T, Boes-Fabian B, Kloetzel PM Koszinowski U (1995) The cleavage preference of the proteasome governs the yield of antigenic peptides. J Exp Med 182:1865–1870

Eleuteri AM, Kohanski RA, Cardozo C, Orlowski M (1997) Bovine spleen multicatalytic proteinase complex (proteasome): replacement of X, Y, and Z subunits by LMP7; LMP2 and MECL1 and changes in properties and specificity. J Biol Chem 272:11824–11831

Enenkel C, Lehmann A, Kloetzel PM (1998) Subcellular distribution of proteasomes implicates a major location of protein degradation in the nuclear envelope-ER network in yeast. EMBO J 17:6144–6154

Falk K, Rotzschke O, Deres K, Metzger J, Jung G, Rammensee HG (1991) Identification of naturally processed viral nonapeptides allows their quantification in infected cells and suggests an allele-specific T cell epitope forecast. J Exp Med 174:425–434

Ferrell K, Wilkinson CRM, Dubiel W, Gordon C (2000) Regulatory subunit interactions of the 26S proteasome, a complex problem. Trends Biochem Sci 25:83–88

Fischer M, Runkel L, Schaller H (1995) HBx protein of hepatitis B virus interacts with the C-terminal portion of a novel human proteasome α-subunit. Virus Genes 10:99–102

Fleming JA, Lightcap ES, Sadis S, Thoroddsen V, Bulawa CE, Blackman RK (2002) Complementary whole-genome technologies reveal the cellular response to proteasome inhibition by PS-341. Proc Natl Acad Sci U S A 99:1461–1466

Frentzel S, Pesold-Hurt B, Seelig A, Kloetzel PM (1994) 20 S proteasomes are assembled via distinct precursor complexes. Processing of LMP2 and LMP7 proproteins takes place in 13–16 S preproteasome complexes. J Mol Biol 236:975–981

Gaczynska M, Rock KL and Goldberg AL (1993) Gamma interferon and expression of MHC genes regulate peptide hydrolysis. Nature 363:262–264

Georgopoulos NT, Proffitt JL, Blair GE (2000) Transcriptional regulation of the major histocompatibility complex (MHC) class I heavy chain, TAP1 and LMP2 genes by the human papillomavirus (HPV) type 6b, 16 and 18 E7 oncoproteins. Oncogene 19:4930–4935

Gerards WL, Enzlin J, Haner M, Hendriks IL, Aebi U, Bloemendal H, Boelens W (1997) The human α-type proteasomal subunit HsC8 forms a double ring-like structure, but does not assemble into proteasome-like particles with the β-type subunits HsDelta or HsBPROS26. J Biol Chem 272:10080–10086

Gerards WLH, deJong, WW, Bloemendal H, Boelens W (1998) The human proteasomal subunit HsC8 induces ring formation of other α-type subunits, J Mol Biol 275:113–121

Glickman MH, Rubin DM, Coux O, Wefes I, Pfeifer G, Cjeka Z, Baumeister W, Fried VA, Finley D. (1998) A subcomplex of the proteasome regulatory particle required for ubiquitin-conjugate degradation and related to the COP9-signalosome and eIF3. Cell 94:615–623

Glickman MH, Rubin DM, Fu H, Larsen CN, Coux O, Wefes I, Pfeifer G, Cjeka Z, Vierstra R, Baumeister W, Fried V, Finley D (1999) Functional analysis of the proteasome regulatory particle. Mol Biol Rep 26:21–28

Glickman MH, Ciechanover A (2002) The ubiquitin-proteasome proteolytic pathway: destruction for the sake of construction. Physiol Rev 82:373–428

Glynne R, Powis S, Beck S, Kelly A, Kerr LA, Trowsdale J (1991) A proteasome related gene between the two abc transporter loci in the calls II region of the human MHC. Nature 353:357–360

Goldberg AL, Cascio P, Saric T, Rock KL (2002) The importance of the proteasome and subsequent proteolytic steps in the generation of antigenic peptides. Mol Immunol 39:147–164

Gottesman S (1996) Proteases and their targets in Escherichia coli. Annu. Rev. Genet 30:465–506

Griffin TA, Nandi D, Cruz M, Fehling HJ, Kaer LV, Monaco JJ, Colbert RA (1998) Immuno-proteasome assembly: cooperative incorporation of interferon-γ (IFN-γ)-inducible subunits. J Exp Med 18:97–104

Griffin, TA, Slack JP, McCluskey TS, Monaco JJ, Colbert RA (2000) Identification of proteassemblin, a mammalian homologue of the yeast protein, Ump1p, that is required for normal proteasome assembly. Mol Cell Biol Res Comm 3:212–217

Groettrup M, Soza M, Eggers M, Kuehn L, Dick TP, Schild HJ Rammensee HG, Koszinowski UH, Kloetzel PM (1996) A role for the proteasome regulator PA28α in antigen presentation. Nature 381:166–168

Groettrup M, Standera, S, Stohwasser R, Kloetzel PM (1997) The subunits MECL-1 and LMP2 are mutually required for incorporation into the 20S proteasome. Proc Natl Acad Sci USA 94:8970–8975

Groll M, Ditzel L, Löwe J, Stock D, Bochtler M, Bartunik HD, Huber R (1997) Structure of 20S proteasome from yeast at 2.4 A resolution. Nature 386:463–471

Groll M, Bakorke M, Kohler A, Moroder L, Rubin DM, Huber R, Glickman M, Finley D (2000) A gated channel into the proteasome core particle. Nature Struct Biol 7:1062–1067

Harding CV (1996) Class I MHC presentation of exogenous antigens. J Clin Immunol 16:90–96

Heinemeyer W, Fischer M, Krimmer T, Stachon U, Wolf DH (1997) The active sites of the eukaryotic 20S proteasome and their involvement in subunit precursor processing. J Biol Chem 272:25200–25209.

Hendil KB, Khan S, Tanaka K (1998). Simultaneous binding of PA28 and PA700 activators to 20S proteasomes. Biochem J 332:749–754

Hill CP, Masters EI, Whitby FG (2002) The 11S regulators of 20S proteasome activity. Curr Top Microbiol Immunol 268:73–89

Holzhütter H-G, P-M Kloetzel (2000) A kinetic model of vertebrate 20S proteasome accounting for the generation of major proteolytic fragments from oligomeric peptide substrates. Biophys J 79:1196–1205

Huang J, Kwong J, Sun EC, Liasng TJ (1996) Proteasome complex as a potential cellular target for hepatitis B virus X protein. J Virol 70:5582–5591

Huang X, Seifert U, Salzmann U, Henklein P, Preissner R, Henke W, Sijts A, Kloetzel PM, Dubiel W (2002) The RTP site shared by the HIV-1 TAT protein and the 11 regulator subunit a is crucial for their effects on proteasome function including antigen processing. J Mol Biol 323:771–782

Jäger S, Groll M, Huber R, Wolf DH, Heinemeyer W (1999) Proteasome β-type subunits: unequal roles of propeptides in core particle maturation and a hierarchy of active site function. J Mol Biol 29:997–1013

Jamaluddin M, Wang S, Garofalo RP, Elliott T, Casola A, Baron S, Brasier AR (2001) IFN-β mediates coordinate expression of antigen-processing genes in RSV-infected pulmonary epithelial cells. Am J Physiol Lung Cell Mol Physiol 280:248–257

Jensen LJ, Knudsen S (2000) Automatic discovery of regulatory patterns in promoter regions based on whole cell expression data and functional annotation. Bioinformatics 16:326–333

Johnson DG, Schneider-Broussard R (1998) Role of E2F in cell cycle control and cancer Front Biosci 27:447–448

Kelly A, Powis SH, Glynne R and Trowsdale J (1991) Second proteasome-related gene in the human MHC class II region. Nature 353:667–668

Kessler JH, Beekman NJ, Bres Vloemans SA, Verdijk P, Van Veelen PA, Kloosterman-Joosten AM, Vissers DC, ten Bosch GJ, Kester MG, Sijts A, Drijfhout JW, Ossendorp F, Offringa R, Melief CJ (2001) Efficient identification of novel HLA-A(*)0201-presented cytotoxic T lymphocyte epitopes in the widely expressed tumor antigen by proteasome-mediated digestion analysis. J Exp Med 193:73–88

Kingsbury DJ, Griffin TA, Colbert RA (2000) Novel propeptide function in 20 S proteasome assembly influences β subunit composition. J Biol Chem 275:24156–162

Kisselev AF, Akopian TN, Castillo V, Goldberg AL (1999) Proteasome active sites allosterically regulate each other, suggesting a cyclical bite-chew mechanism for protein breakdown. Mol Cell 4:395–402

Kisselev AF, Akopian TN, Goldberg AL (1998). Range of sizes of peptide products generated during degradation of different proteins by archeal proteasomes. J Biol Chem 274:3363–3371

Kisselev AF, Kaganovich D, Goldberg AL. (2002) Binding of hydrophobic peptides to several noncatalytic sites promotes peptide hydrolysis by all active sites of 20 S proteasomes. Evidence for peptide-induced channel opening in the α-rings. J Biol Chem 277:22260–22270

Kloetzel PM (2001) Antigen processing by the proteasome. Nat Rev Mol Cell Biol 2:179–187

Knuehl C, Seelig A, Brecht B, Henklein P, Kloetzel PM (1996) Functional analysis of eukaryotic 20S proteasome nuclear localization signal. Exp Cell Res 225 67–74

Knuehl C, Spee P, Ruppert T, Kuckelkorn U, Henklein P, Neefjes J, Kloetzel PM (2001) The murine cytomegalovirus pp89 immunodominant H-2Ld epitope is generated and translocated into the endoplasmic reticulum as an 11-mer precursor peptide. J Immunol 167:1515–1521

Knowlton JR, Johnston SC, Whitby FG, Realini C, Zhang Z, Rechsteiner M, Hill CP (1997) Structure of the proteasome activator REGα (PA28α). Nature 390:639–643

Köhler A, Cascio P, Legget DS, Woo KM, Goldberg AL, Finley D (2001) The axial channel of the proteasome core particle is gated by the rpt2 ATPase and controls both substrate entry and product release. Mol Cell 7:1143–1152

Kopp F, Dahlmann B, Kuehn L (2001) Reconstitution of hybrid proteasomes from purified PA700–20 S complexes and PA28$\alpha\beta$ activator: ultrastructure and peptidase activities. J Mol Biol 313:465–471

Krüger E, Zühlke D, Witt E, Ludwig H, Hecker M (2001) Clp-mediated proteolysis in Gram-positive bacteria is autoregulated by the stability of a repressor. EMBO J 20:852–863

Krüger E, Hecker M (1998) The first gene of the *clpC*-operon in *Bacillus subtilis, ctsR* encodes a negative regulator of its own operon and other class III heat shock genes. J Bacteriol 180:6681–6688

Krüger E, Msadek T, Hecker M (1996) Alternate promoters direct stress induced transcription the of *Bacillus subtilis clpC* operon. Mol Microbiol 20:713–723

Krüger E, Kloetzel PM, Enenkel C (2001) 20S proteasome biogenesis. Biochimie 83:289–294

Kuckelkorn U, Frentzel S, Kraft R, Kostka S, Groettrup M, Kloetzel PM (1995) Incorporation of major histocompatibility complex encoded subunits LMP2 and LMP7 changes the quality of the 20S proteasome polypeptide processing products independent of interferon-γ. Eur J Immunol 25:2605–2611

Kuckelkorn U, Ruppert T, Strehl B, Jungblut PR, Zimny-Arndt U, Lamer S, Prinz I, Drung I, Kloetzel PM, Kaufmann SH, Steinhoff U (2002a) Link between organ-specific antigen processing by 20S proteasomes and CD8(+) T cell-mediated autoimmunity. J Exp Med 195:983–990

Kuckelkorn U, Ferreira EA, Drung I, Liewer U, Kloetzel PM, Theobald M (2002b) The effect of the interferon-gamma-inducible processing machinery on the generation of a naturally tumor-associated human cytotoxic T lymphocyte epitope within a wild-type and mutant p53 sequence context. Eur J Immunol 32:1368–1375

Kuon W, Holzhutter HG, Appel H, Grolms M, Knollnberger S, Traeder A, Henklein P, Weiss E, Thiel A, Lauster R, Bowness P, Radbruch A, Kloetzel PM and Sieper J (2001) Identification of HLA-B27-restricted peptides from *Chlamydia trachomatis* proteome with possible relevance to HLA-B27-associated diseases. J Immunol 167:4738–4746

Kuttler C, Nussbaum AK, Dich TP, Rammensee HG, Schild H, Hadeler KP (2001) An algorithm for the prediction of proteasome cleavage. J Mol Biol 298: 417–429

Lehmann A, Janek K, Braun B, Kloetzel PM, Enenkel C (2002) 20S proteasomes are imported as precursor complexes into the nucleus of yeast. J Mol Biol 317:401–413

Levitskaya J, Coram M, Levitsky V, Imreh S, Steigerwald-Mullen PM, Klein G, Kurilla MG, Masucci MG (1995). Inhibition of antigen processing by the internal repeat region of the Epstein-Barr virus nuclear antigen-1. Nature 375:685–688

Levitskaya J, Sharipo A, Leonchiks A, Ciechanover A, Masucci MG (1997) Inhibition of ubiquitin/proteasome-dependent protein degradation by the Gly-Ala repeat domain of the Epstein-Barr virus nuclear antigen 1. Proc Natl Acad Sci U S A 94:12616–12621

Levy F, Burri L, Morel S, Peitrequin AL, Levy N, Bachi A, Hellman U, Van den Eynde BJ, Servis C (2002) The final N-terminal trimming of a subaminoterminal proline-containing HLA class I-restricted antigenic peptide in the cytosol is mediated by two peptidases. J Immunol 169:4161–4171

Li J, Gao X, Ortega J, Nazif T, Joss L, Bogyo M, Steven AC, Rechsteiner M (2001) Lysine 188 substitutions convert the pattern of proteasome activation by REGγ to that of REGs α and β. EMBO J 20:3359–3369

Löwe J, Stock D, Jap B, Zwickl P, Baumeister W, Huber R (1995) Crystal structure of the 20S proteasome from the archaeon T. acidophilum at 3.4 A resolution. Science 268:533–539

Ma CP, Slaugther CA, DeMartino GN (1992) Identification, purification and characterisation of a protein activator (PA28) of the 20S proteasome (macropain). J Biol Chem 267:10115–10123

Macagno A, Gillet M, Sallust F, Lanzavecchia A, Nestle FO, Groettrup M (1999) Dendritic cells upregulate immunoproteasomes and the proteasome regulator PA28 during maturation. Eur J Immunol 29:4037–4042

Madden DR (1995) The three-dimensional structure of peptide-MHC complexes. Annu Rev Immunol 13:587–622

Mannhaupt G, Schnall R, Karpov V, Vetter I, Feldmann H (1999) Rpn4p acts as a transcription factor by binding to PACE, a nonamer box found upstream of 26S proteasomal and other genes in yeast. FEBS Lett 450:27–34

Martinez CK and Monaco JJ (1991) Homology of proteasome subunits to major histocompatibility complex linked LMP gene. Nature 353:664–667

Mayr J, Seemuller E, Muller SA, Engel A, Baumeister W (1998) Late events in the assembly of 20S proteasomes. J Struct Biol 124:179–188

McCutchen-Maloney SL, Matsuda K, Shimbara N, Binns DD, Tanaka K, Slaughter CA, DeMartino GN (2000) cDNA cloning, expression, and functional characterization of PI31, a proline-rich inhibitor of the proteasome. J Biol Chem 275:18557–18565

Michalek MT, Grant EP, Gramm C, Goldberg AL, Rock KL (1993) A role for the ubiquitin-dependent proteolytic pathway in MHC class I restricted antigen presentation. Nature 363:552-554

Missiakas D, Schwager F, Betton JM, Georgopolus C, Raina S (1996) Identification and characterization of HslV HslU (ClpQ ClpY) proteins involved in overall proteolysis of misfolded proteins in *Escherichia coli*. EMBO J 15:6899–6909

Momburg F, Ortiz-Navarrete V, Neefjes J, Goulmy E, van de Wal Y, Spits H, Powis SJ, Butcher GW, Howard JC, Walden P (1992) Proteasome subunits encoded by the major histocompatibility complex are not essential for antigen presentation. Nature 360:174–177

Morel S, Levy F, Burleit-Schiltz O, Brasseur F, Probst-Kepper M, Peitrequin AL, Monsarrat B, Van Velthoven R, Certtini JC, Boon T, Gairin JE, Van den Eynde BJ (2000) Processing of some antigens by standard proteasome but not by the immunoproteasome result in poor presentation by dendritic cells. Immunity 12:101–117

Murata S, Udono H, Tanahashi N, Hamada N, Watanabe K, Adachi K, Yamano T, Yui K, Kobayashi N, Kasahara M, Tanaka K, Chiba T (2001) Immunoproteasome assembly and antigen presentation in mice lacking both PA28α and PA28β. EMBO J 20:5898–5907

Nandi D, Jiang H, Monaco JJ (1996) Identification of MECL-1 (LMP-10) as the third IFN-γ-inducible proteasome subunit. J Immunol 156:2361–2364

Nandi D, Woodward E, Ginsburg DB, Monaco JJ (1997) Intermediates in the formation of mouse 20S proteasomes: implications for the assembly of precursor subunits. EMBO J 16:5363–5375

Niedermann G, Grimm R, Geier E, Maurer M, Realini C, Gartmann C, Soll J, Omura S, Rechsteiner MC, Baumeister W, Eichmann K (1997) Potential immunocompetence of proteolytic fragments produced by proteasomes before evolution of the vertebrate immune system. J Exp Med 186:209–220

Nussbaum AK, Dick TP, Keilholz W, Schirle M, Stevanovic S, Dietz K, Heinemeyer W, Groll M, Wolf DH, Huber R, Rammensee HG, Schild H (1998) Cleavage motifs of the yeast proteasome β subunits deduced from digests of enolase 1. Proc Natl Acad Sci USA 95:12504–12509

Orlowski M, Cardozo C, Michaud C (1993) Evidence for the presence of five distinct proteolytic components in the pituitary multicatalytic proteinase complex. Properties of two components cleaving bonds on the carboxyl side of branched chain and small neutral amino acids. Biochemistry 32:1563–1572

Ortiz-Narvarete V, Seelig A, Gernold M, Frentzel S, Kloetzel PM, Hämmerling GJ (1991) Subunit of the 20S proteasome (multicatalytic proteinase) encoded by the major histocompatibility complex. Nature 353:662–672

Ossendorp F, Eggers M, Neisig A, Ruppert T, Groettrup M, Sijts A, Mengede E, Kloetzel PM, Neefjes J, Koszinowski U, Melief C (1996) A single residue exchange within a viral CTL epitope alters proteasome-mediated degradation resulting in lack of antigen presentation. Immunity 5:115–124

Palmer A, Mason GG, Paramio JM, Knecht E, Rivett AJ (1994) Changes in proteasome localization during the cell cycle. Eur J Cell Biol 64:163–175

Palmer A, Rivett AJ, Thomson S, Hendil KB, Butcher GW, Fuertes G, Knecht E (1996) Subpopulations of proteasomes in rat liver nuclei, microsomes, and cytosol. Biochem J 316:401–407

Peters JM, Cejka Z, Harris JR, Kleinschmidt JA, Baumeister W (1993) Structural features of the 26 S proteasome complex.J Mol Biol 234:932–937

Peters JM, Franke WW, Kleinschmidt JA (1994) Distinct 19S and 20S subcomplexes of the 26S proteasome and their distribution in the nucleus and the cytoplasm. J Biol Chem 269:7709–7718

Plemper RK, Wolf DH (1999) Retrograde transport translocation: Reverse protein transport and its end in the proteasome. Trends Biochem Sci 24: 266–270

Preckel T, Fung-Leung WP, Cai Z, Vitiello A, Salter-Cid L, Winqvist O, Wolfe TG, Von Herrath M, Angulo A, Ghazal P, Lee J, Fourie AM, Wu Y, Pang J, Ngo K, Peterson PA, Früh K, Yang Y (1999) Impaired immunoproteasome assembly and immune response in PA28-/- mice. Science 286:2162–2165

Proffitt JA, Blair GE (1997) The MHC-encoded TAP1/LMP2 bidirectional promoter is downregulated in highly oncogenic adenovirus type 12 transformed cells. FEBS Lett 400:141–144

Ramos PC, Höckendorff J, Johnson ES, Varshavsky A, Dohmen RJ (1998) Ump1p is required for proper maturation of the 20S proteasome and becomes its substrate upon completion of the assembly. Cell 92:489–499

Reits EA, Benham AM, Plougastel B, Neefjes J, Trowsdale J (1997) Dynamics of proteasome distribution in living cells. EMBO J 16:6087–6094

Restifo NP, Esquivel F, Kawakami Y, Yewdell JW, Mule JJ, Rosenberg SA, Bennink JR (1993) Identification of human cancers deficient in antigen processing. J Exp Med 177:265–272

Rock KL, Gramm C, Rothstein L, Clark K, Stein R, Dick L, Hwang D, Goldberg AL (1994) Inhibitors of the proteasome block the degradation of most cell proteins and the generation of peptides presented on MHC class I molecules. Cell 78:761–771

Rock LK, York IA, Saric T, Goldberg AL (2002) Protein degradation and the generation of MHC class I-presented peptides. Adv Immunol 80:1–59

Rotem-Yehudar A, Groettrup M, Soza A, Kloetzel PM, Ehrlich R (1996) LMP-associated proteolytic activities and TAP-dependent peptide transport for class I MHC molecules are suppressed in cell lines transformed by the highly oncogenic adenovirus 12. J Exp Med 183:499–514

Rousset R, Desbois C, Bantignies F, Jalinot P (1996) Effects of the NF-kB1/105 processing of the interaction between HTLV-1 transactivator Tax and the proteasome. Nature 381:328–331

Salzmann U, Kral S, Braun B, Standera S, Schmidt M, Kloetzel PM, Sijts A (1999) Mutational analysis of subunit i β2 (MECL-1) demonstrates conservation of cleavage specificity between yeast and mammalian proteasomes. FEBS Lett 54:11–15

Saric T, Chang SH, Hattori A, York IA, Markant S, Rock KL, Tsujimoto M, Goldberg AL (2002) ERAP1, an interferon-γ-induced aminopeptidase in the endoplasmic reticulum, that trims precursors to MHC class I-presented peptides. Nature Immunol (in press)

Saveanu L, Fruci D, van Endert P (2002) Beyond the proteasome: trimming, degradation and generation of MHC class I ligands by auxiliary proteases. Mol Immunol 39:203–214

Scheffner M, Hulbregtse JM, Vierstra RD, Howley PM (1993) The HPV-16 E6 and E6-AP complex functions as a ubiquitin-protein ligase in the ubiquitination of p53. Cell 75:495–505

Schmidt M, Zantopf D, Kraft R, Kostka S, Preissner R, Kloetzel PM (1999) Sequence information within proteasomal prosequences mediates efficient integration of β-subunits into the 20 S proteasome complex. J Mol Biol 288:117–128

Schmidtke G, Kraft R, Kostka S, Henklein P, Frömmel C, Löwe J, Huber R, Kloetzel PM, Schmidt M (1996) Analysis of mammalian 20S proteasome biogenesis: The maturation of β-subunits is an ordered two-step mechanism involving autocatalysis. EMBO J 15:6887–6898

Schmidtke G, Schmidt M, Kloetzel PM (1997) Maturation of mammalian 20 S proteasomes: Purification and characterization of proteasome precursor complexes. J Mol Biol 268:95–106

Schmidtke G, Eggers M, Ruppert T, Groettrup M, Koszinowski U, Kloetzel PM (1998) Inactivation of a defined active site in the mouse 20S proteasome complexes enhances MHC class I presentation of a murine cytomegalovirus protein. J Exp Med 187:1641–1646

Schmidtke G, Emch S, Groettrup M, Holzhütter HG (2000) Evidence for the existence of a noncatalytic modifier site of peptide hydrolysis by the 20S proteasome. J Biol Chem 275:22056–22063

Schubert U, Antón LC, Gibbs J, Norbury CC, Yewdell JW, Bennink (2000) Rapid degradation of a large fraction of newly synthesised proteins by proteasomes. Nature 404:770–774

Schwarz K, Eggers M, Soza A, Koszinowski UH Kloetzel PM, Groettrup M (2000a) The proteasome regulator PA28α/β can enhance antigen presentation without affecting 20S proteasome subunit composition. Eur J Immunol 165:768–778

Schwarz, K, van Den Broek M, Kostka S, Kraft R, Soza A, Schmidtke G, Kloetzel PM, Groettrup M (2000b) Overexpression of the proteasome subunits LMP2, LMP7, and MECL-1 but not PA28α/βenhances the presentation of an immunodominant lymphocyte choriomeningitis virus T cell epitope. J Immunol 165:768–778

Seeger M, Ferrell K, Frank R, Dubiel W (1997) HIV-tat inhibits the 20S proteasome and its 11S regulator-mediated activation. J Biol Chem 272:8145–8148

Seliger B, Hohne A, Knuth A, Bernhard H, Meyer T, Tampe R, Momburg F, Huber C (1996) Analysis of the major histocompatibility complex class I antigen presentation machinery in normal and malignant renal cells: evidence for deficiencies associated with transformation and progression. Cancer Res 56:1756–1760

Seliger B, Wollscheid U, Momburg F, Blankenstein T, Huber C (2001) Characterization of the major histocompatibility complex class I deficiencies in B16 melanoma cells. Cancer Res 61:1095–1099

Serwold T, Gonzalez F, Kim J, Jacob R, Shastri N (2002) ERAAP customizes peptides for MHC class I molecules in the endoplasmic reticulum. Nature 419:480–483

Shimbara N, Nakajama H, Tanahasi N, Ogawa K, Niwa S, Uenakla A, Nakayama E, Tanaka K (1997) Double cleavage production of the CTL epitope by proteasomes and PA28: role of the flanking region. Genes Cells 2:785–800

Sijts A, Ruppert T, Rehermann B, Schmidt M., Koszinowski U, Kloetzel PM (2000a). Structural features of immunoproteasomes determine the efficient generation of a viral CTL epitope. J Exp Med 191:503–513

Sijts A, Standera S, Beekmann N, van Veelen P-A, Ossendorp FA, Melief CJ, Kloetzel PM (2000b) MHC class I antigen processing of an Adenovirus CTL epitope is linked to the levels of immunoproteasomes in infected cells. J Immunol 164:450–456

Sijts A, Sun Y, Janek K, Paschen A, Schadendorf D, Kloetzel PM (2002) The role of proteasome activator PA28 in antigen processing. Mol Immunol 39:165–172

Soza A, Knuehl C, Groettrup M, Henklein P, Tanaka K, Kloetzel PM (1997) Expression and subcellular localization of mouse 20S proteasome activator complex PA28. FEBS Lett 413:27–34

Stohwasser R, Standera S, Peters I, Kloetzel PM, Groettrup M (1997) Molecular cloning of the mouse proteasome subunits MC14 and MECL-1: reciprocally regulated tissue expression of interferon-γ-modulated proteasome subunits. Eur J Immunol 27:1182–1187

Stohwasser R, Soza A, Eggers M, Koszinowski U, Kloetzel PM (2000a) PA28α/β double and PA28β single transfecttant mouse B8 cell lines reveal enhanced presentation of a mouse cytomegalovirus (MCMV) pp89 MHC class I epitope. Mol Immunol 37:13–19

Stohwasser R, Salzmann U, Giesebrecht J, Kloetzel PM, Holzhütter HG (2000b) Kinetic evidences for facilitation of peptide channeling by the proteasome activator PA28. Eur J Biochem 276:6221–6229

Stohwasser R, Holzhütter HG, Lehmann U, Henklein P, Kloetzel PM (2003) Hepatitis B virus HBx peptide 116–138 and proteasome activator PA28 compete for binding to the proteasome α4/MC6 subunit. Biol Chem 384:39–49

Stoltze L, Schirle M, Schwarz G, Schroter C, Thompson MW, Hersh L, Kalbacher H, Stevanovich S, Rammensee HG, Schild HJ (2000) Two new proteases in the MHC class I processing pathway. Nat Immunol 1:413–418

Strickland E, Hakala K, Thomas PJ, DeMartino GN. (2000) Recognition of misfolded proteins by PA 700, the regulatory subcomplex of the 26S proteasome. J Biol Chem 275:5565–5572

Sun Y, Sijts A, Song M, Janek K, Nussbaum AK, Kral S, Schirle M, Stevanovic S, Paschen A, Schild HJ, Kloetzel PM, Schadendorf D (2002) Expression of the proteasome activator PA28 rescues the presentation of a cytotoxic T lymphocyte epitope on melanoma cells. Cancer Res 62:2875–2882

Tanahashi N, Murakami Y, Minami Y, Shimbara N, Hendil KB, Tanaka K (2000) Hybrid proteasomes: Induction by interferon-χ and contribution to ATP-dependent proteolysis. J Biol Chem 275:14336–14345

Toes RE, Nussbaum AK, Degermann S, Schirle M, Emmerich NP, Kraft M, Laplace C, Zwindemann A, Dicj TP, Muller J, Schonfisch B, Schmid C, Fehling HJ, Stevanovic S, Rammensee, HG, Schild HJ (2001) Discrete cleavage motifs of constitutive and immunoproteasomes revealed by quantitative analysis of cleavage products. J Exp Med 194:1-12

Turnell AS, Grand RJA, Gorbea C, Zhang X, Wang W, Mymryk JS, Gallimore PH (2000) Regulation of 26S proteasome by adenovirus. EMBO J 19:4759–4773

Ustrell V, Pratt G and Rechsteiner M (1995) Effects of interferon $\gamma$and major histocompatibility complex encoded subunits on peptidase activities of human multicatalytic proteases. Proc Natl Acad Sci USA 92:584–588

van Hall T. Sijts A, Camps M, Offringa R, Melief C, Kloetzel PM, Ossendorp F (2000) Differential influence on CTL epitope presentation by controlled expression of either proteasome immuno-subunits or PA28. J Exp Med 192:483–492

Verma R, Chen S, Feldman R, Schieltz D, Yates J, Dohmen J, Deshaies RJ (2000) Proteasomal proteomics: identification of nucleotide-sensitive proteasome-interacting proteins by mass spectrometric analysis of affinity-purified proteasomes. Mol Biol Cell 11:3425–3439

Voges D, Zwickl P, Baumeister W (1999) The 26S proteasome: A molecular machine designed for controlled proteolysis. Annu Rev Biochem 68:1015–1068

Whitby FG, Master EI, Kramer L, Knowlton JR, Yao Y, Wang CC, Hill CP (2000) Structural basis for the activation of 20S proteasomes by 11S regulators. Nature 408:115–120

Wilk S and Orlowski M (1983) Evidence that pituitary cation-sensitive neutral endopeptidase is a multicatalytic protease complex. J Neurochem 40:842–849

Witt E, Zantopf D, Schmidt M, Kraft R, Kloetzel PM, Krüger E (2000) Characterization of the newly identified human Ump1 homologue POMP and analysis of LMP7 (β5i) incorporation into 20S proteasomes. J Mol Biol 301:1-9

Wojcik C, Paweletz N, Schroeter D (1995) Localization of proteasomal antigens during different phases of the cell cycle in HeLa cells. Eur J Cell Biol 68:191–198

Xie Y, Varshavsky A (2001) RPN4 is a ligand, substrate, and transcriptional regulator of the 26S proteasome: a negative feedback circuit. Proc Natl Acad Sci U S A 98:3056–3061

Yang Y, Fruh K, Ahn K, Peterson PA (1995) In vivo assembly of the proteasomal complexes, implications for antigen processing. J Biol Chem 270:27687–27694

Yao Y, Huang L, Krutchinsky A, Wong ML, Standing KG, Burlingame AL, Wang CC (1999) Structural and functional characterizations of the proteasome-activating protein PA26 from *Trypanosoma brucei*. J Biol Chem 274:33921–33930

Yewdell J, Lapham C, Bacik I, Spies T, Bennink J (1994) MHC-encoded proteasome subunits LMP2 and LMP7 are not required for efficient antigen presentation. J Immunol 152:1163–1170

Zaiss DM, Standera S, Holzhutter H, Kloetzel P, Sijts AJ (1999) The proteasome inhibitor PI31 competes with PA28 for binding to 20S proteasomes. FEBS Lett 457:333–338

Zaiss D, Standera S, Kloetzel PM and Sijts A (2002) PI31 is a modulator of proteasome formation and antigen processing. Proc Natl Acad Sci U S A 99:14344–14349

Zühl F, Seemüller E, Golbik R, Baumeister W (1997) Dissecting the assembly pathway of the 20S proteasome. FEBS Lett 418:189–194

Zwickl P, Kleinz J, Baumeister W (1994) Critical elements in proteasome assembly. Nat Struct Biol 11:765–770

Rev Physiol Biochem Pharmacol (2003) 148:105–150
DOI 10.1007/s10254-003-0011-3

G. Fischer · T. Aumüller

Regulation of peptide bond *cis/trans* isomerization by enzyme catalysis and its implication in physiological processes

Published online: 16 April 2003

Abstract In some cases, the slow rotational movement underlying peptide bond *cis/trans* isomerizations is found to control the biological activity of proteins. Peptide bond *cis/trans* isomerases as cyclophilins, Fk506-binding proteins, parvulins, and bacterial hsp70 generally assist in the interconversion of the polypeptide substrate *cis/trans* isomers, and rate acceleration is the dominating mechanism of action in cells. We present evidence disputing the hypothesis that some of the molecular properties of these proteins play an auxiliary role in enzyme function.

Introduction

In this review, discussion is focused on the recent progress in understanding peptide bond *cis/trans* isomerase function in cells, and the help they can give other proteins in playing their biological role. It is now clear that the spatial and temporal characteristics of protein folding dictate most aspects of cell life. Because of the involvement of peptide bond *cis/trans* interconversions in protein folding, and because peptide bonds will adopt relatively stable nonnative conformations during folding, the existence of folding intermediates and folding pathways needs to be considered in a comprehensive treatment of folding helper proteins.

It is also clear that, when a polypeptide binds to a native protein, the resulting complex forms a global ensemble that must seek its own minimum free energy fold. Consequently, the rules underlying the process of de novo protein folding do not change, in principle for chain rearrangements induced in native proteins.

It is rather obvious that for isolated polypeptide chains to get properly folded or even misfolded a huge number of conformational changes on a potential energy landscape are required that also involve an enormous variety of energy minima and pathways connecting

G. Fischer (✉) · T. Aumüller
Max Planck Research Unit for Enzymology of Protein Folding,
Weinbergweg 22, 06120 Halle, Germany
e-mail: fischer@enzyme-halle.mpg.de · Tel.: +49-345-5522800 · Fax: +49-345-5511972

them (Dobson et al. 1998). Under conditions reflecting the intracellular situation, complexity is further increased due to macromolecular crowding (Frydman 2001). Thus, devising models for protein folding on the submolecular level and predicting folding pathways is often found to be an unrewarding task.

Interpretation of a folding reaction becomes much easier when we restrict our attention to a conformational interconversion that confers uncoupling of some bond rotations from the framework movement of the backbone and side chain packing by virtue of exceptionally high torsional barriers.

Usually, the energy barrier separating these conformers will be high enough that interconversion occurs on a time scale ranging from the upper millisecond range to a few hours. Now suppose that in a folding reaction or an intermolecular protein-ligand recognition process only one of the coexisting conformers is compatible with the final low-energy structure. Then, the fraction of the conformer not readily recognized is left over for a certain period of time. This in turn must imply the inability of an induced fit mechanism to generate promiscuous binding of the complete ensemble of high-energy barrier conformers existing free in solution. Thus, conformer-specific reaction kinetics can be observed, characterized by a transient change in kinetic parameters during reaction completion. The height of the activation energy barrier determines the time-scale of the transients, with a low barrier responsible for fast transients. Consequently, in various folding states of a polypeptide chain one or more covalent bonds can adopt metastable conformations susceptible to kinetic trapping. Intriguingly, this is an example for uncoupled folding that might allow for the tracing a particular bond during the reaction progress. The idea that conformational changes at a discrete bond report on folding progress becomes apparent at monitoring slow folding phases in the time-course of refolding in vitro, as could be realized by intelligent serial jump techniques (Schmid 1986).

An important implication is that chain reshuffling by protein/ligand (substrate) interaction, de novo protein folding, and refolding of denatured proteins is based on similar molecular principles (Tsai et al. 1999), while the starting conformational states are very different. Since the pioneering work of Brandts and his coworkers in the mid-1970s (Brandts et al. 1975), knowledge has increased on the slow kinetic refolding phases of chemically denatured proteins and isomer-specific intermolecular biorecognition. Activation energies range from 60 kJ mol^{-1} to 90 kJ mol^{-1}, relatively independent of the experimental conditions, and were assigned to the presence of prolyl bonds in the polypeptide chain. The term prolyl bond used throughout the paper indicates the imidic peptide bond preceding proline in an amino acid sequence, and prolyl isomerization indicates the *cis/trans* isomerization of the peptide bond preceding proline.

Nineteen of the 20 gene-coded amino acids form secondary amide peptide bonds (also termed "normal" or "nonproline" peptide bond), whereas the N-alkylated amino acid proline has the potential to create an imidic peptide bond at its N-terminal tail.

Due to the imidic structure, prolyl bonds have a particular electron distribution and steric constraints. Various folding probes reflect high rotational barriers for a considerable fraction of molecules when monitoring the folding progress, pointing to the involvement of prolyl bonds in the folding reaction. Recent investigations utilizing ^{1}H-NMR approaches (Scherer et al. 1998) and double jump experiments (Odefey et al. 1995) have shown that high rotational barriers exist not only for prolyl bonds but also for secondary amide peptide bonds. Excepting nonprolyl peptide bonds in *cis* conformation in native proteins, which rarely occur, the visible level of amplitudes in refolding probes is greatly reduced

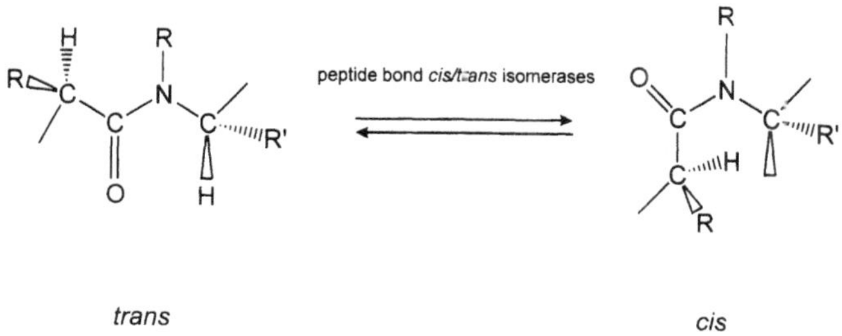

Fig. 1 The *cis/trans* isomers of a peptide bond

Fig. 2 The transition state diagrams for the uncatalyzed and the enzyme-catalyzed peptide bond *cis/trans* isomerization. The half-way position of the transition state between *cis* and *trans* ground states for the uncatalyzed reaction can be shifted for the enzyme-catalyzed reaction. Relative free energies approximate a tetrapeptide situation when bound to Cyp18 (from Kern et al 1995)

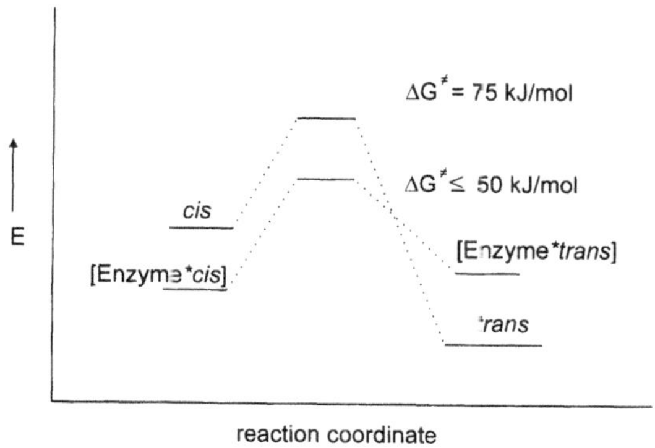

for this isomerizing group, often making detection difficult (Pappenberger et al. 2001). The mechanism by which C-N bonds may confer rotational uncoupling of folding is through the latter's rigidity bordered by freely rotatable single bonds (Fig. 1). In addition, the properties of this bond have important implications for the chain stiffness in all folding states of a polypeptide chain.

For both types, the imide and the secondary amide bonding, the classical view of the amide resonance is based on the ability of the nitrogen atom to delocalize its lone pair over the whole functional group. The resulting planar framework of electronic organization and the consequent partial double-bond character of the C-N bond cause restrictions for the number of energy minima in amide bond torsion (dihedral angle ω). The free energy dependence of the reaction progress of the peptide bond *cis/trans* isomerization in a chain fragment indicates just two minimum-energy structures. The geometric isomers at the angles $\omega \approx 0°$ (*cis*) and $\omega \approx 180°$ (*trans*) are separated by the rotational barrier corresponding to the perpendicular high-energy state of $\omega \approx 90°$ (Fig. 2). The distinction between the *cis* and *trans* isomers of a molecule originates from a geometry-based classification of organic molecules.

In oligopeptides and unfolded polypeptides, the general preference of secondary amide peptide bonds for the *trans* conformation with an equilibrium constant K=[*cis*]/[*trans*] = $(k_{trans\ \mathrm{to}\ cis} / k_{cis\ \mathrm{to}\ trans}) << 10^{-2}$ (Scherer et al. 1998; Schiene-Fischer and Fischer 2001), where k (s^{-1}) stands for the rate constants of the *cis/trans* interconversion, is somewhat reduced for prolyl bonds. Depending on the amino acids flanking the proline residue, K is in the range of 0.1–1.0 (Yao et al. 1994; Reimer et al. 1998).

In contrast, three-dimensional protein structures from crystals reveal a defined isomeric state for every prolyl peptide bond being either in *cis* or *trans* conformation, with the majority of prolyl bonds in *trans* (MacArthur and Thornton 1991; Reimer and Fischer 2002). The seeming homogeneity only partially results from isomer-specific stabilization effects by intramolecular long-range interactions. When probed by NMR spectroscopy in solution, many proteinaceous prolyl bonds show conformational polymorphism, resulting in a wide range of isomer ratio values for the native state. Limitations such as compromised ability to ascertain minor conformers and dominating crystal-packing forces must be considered for detection of biologically significant conformers in the crystal structure database.

In terms of rotational dynamics, the peptide bond *cis/trans* isomerization results in an slow oscillation of chain segments between isomers of different end-to-end lengths, with the shorter distance in the *cis* conformation. The oscillating portion of the backbone is differentially extended in length for different proteins (Reimer and Fischer 2002; Li et al. 2001a). In conclusion, the oscillatory properties of the peptide bond *cis/trans* isomerization resemble a molecular switch with the characteristics of (1) being stably arrested in only two positions; (2) the resistance to switching is high on the energy scale of biorecognition; (3) the switch can amplify signals by lever effects; (4) enzymes externally control the switching resistance in vivo (Fischer 1994a; Reimer and Fischer 2002). Coupling a conformational equilibrium of high switching resistance to fast subsequent reaction steps may cause a transient discrimination of conformers in the overall reaction, often termed isomer-specificity.

Proteolysis of proline-containing oligopeptides provides the first evidence that not only the intramolecular polypeptide chain association distinguishes between coexisting conformers during the formation of folding intermediates, but that the intermolecular association of enzymes and oligopeptide substrates does so as well. For enzymes, the Gibbs energy difference of transition state binding following the formation of the Michaelis complexes by ground state binding of substrates determines the power of catalysis. Both ground state and transition state binding to the protease were found to be dependent upon the substrate conformation. Interestingly, conformer-specific effects have been obtained for protease substrates that differ in the backbone conformation only within sequence regions remote to the scissile bond (Fischer et al. 1983, 1984a). Protease-independent kinetic phases result by partitioning substrate conformers into differently reactive Michaelis complexes, where the first-order rate constants of the phases correlate with the magnitude of the peptide bond rotational barriers by the Arrhenius equation.

If the conformational switch reveals a high turning resistance, then kinetic traps of different lifetime exist for folding intermediates, polypeptide ligands, and native proteins. Large proteins might especially suffer from persistent kinetic traps in refolding, deteriorating the productive folding pathway. Experience gained from the conduct of reaction mechanisms suggests that catalysis of the isomerization reaction may be utilized as a means of escaping kinetic traps. However, intermolecular catalysis of peptide bond *cis/trans* isomerization by small molecules proved to be difficult in aqueous solution. Therefore, acceleration of peptide bond *cis/trans* isomerization in cells has to be associated with enzymes that essentially involve an active site cavity and a particular mechanism of catalysis.

With the first enzyme discovered in 1984, this enzyme class represents the unique example of biocatalysts directed toward a conformational interconversion in proteins known to date (Fischer et al. 1984b). Enzymes catalyzing prolyl peptide bond *cis/trans* isomerization evolve early during the course of evolution, and three different enzyme families emerge soon after the progenitor has put in an appearance. According to our current knowledge, the peptidyl prolyl *cis/trans* isomerases (PPIases) comprise cyclophilins (Cyp), FK506 binding proteins (FKBP), and parvulins (Par). The family members are unrelated to each other in their amino acid sequences, have distinct substrate specificities, and prove to be sensitive to different types of inhibitors. Archetypical PPIases are monomerically active enzymes ranging in size from 92 to 163 amino acids. Recently, the secondary amide peptide bond *cis/trans* isomerases (APIases) were added to this class of enzymes, indicating an expanded set of polypeptide substrates sensitive to biocatalysis (Schiene-Fischer et al. 2002a).

Reaction profile of peptide bond *cis/trans* isomerases

Previous biochemical investigations have led to the elucidation of four different molecular mechanisms that might control, either in isolation or collectively, the physiological role of archetypical peptide bond *cis/trans* isomerases:

1. The catalysis of peptide bond *cis/trans* isomerization
2. A holding function for unfolded polypeptide chains
3. A presenter protein function for an unknown number of physiological ligands
4. A proline-directed binding function (for PPIases)

The calcium/magnesium-dependent nuclease activity reported for cyclophilins (Montague et al. 1994) might originate from an impurity present in cyclophilin preparation (Schmidt et al. 1996). Among all known enzymes, peptide bond *cis/trans* isomerases exhibit an unique reaction profile with a great number of potentially reactive sites in a polypeptide substrate, and a low degree of chemical differences between reactand and product.

In a considerable portion of peptide bond *cis/trans* isomerases supplementary domains add to the physiological function according to their individual properties.

Submolecular principles underlying physiological control through catalyzed peptide bond *cis/trans* isomerizations, which represent the mechanism 1, could be: (1a) attenuation of the kinetic barrier underlying isomer-specific recognition of a polymorphic polypeptide backbone or (1b) site-directed stiffness control in the peptide bond C-N linkage.

In view of isomer-specific transition state binding, enzymes involved cannot convert the *cis* and the *trans* isomers of a substrate into products at the same rate. In contrast, discrimination of an isomer in ground state binding can lead to both isomer-specific equilibrium constants and isomer-specific reaction rates (Fig. 2).

Isomer-specific binding can be realized intra- and intermolecularly The detection of type (1a) reactions requires a notable concentration of the biologically less active conformer in its unbound state. On the other hand, the catalysis by a peptide bond *cis/trans* isomerase inevitably involves reduction of the stiffness-determining C-N double-bond character, which in turn leads to an increase of bond rotation rates. In essence, this effect is in accord with the submolecular principle (1b; see above). Importantly, a comparable population of free conformers in solution is not a prerequisite for this effect. It only depends on the monomolecular interconversion rates of the Michaelis complexes. It is evident from the reaction cycle for the enzyme-catalyzed *cis/trans* isomerization that the internal equilibrium constant K_{int} = [E**trans*]/[E**cis*] for the Michaelis complexes exists and

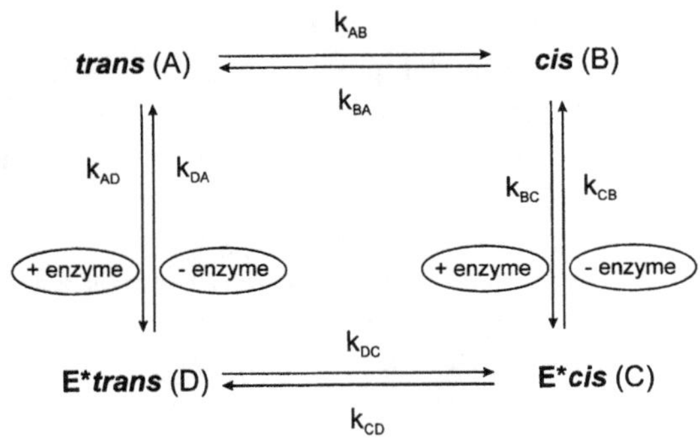

Fig. 3 The minimal reaction scheme for the enzyme-catalyzed peptide bond *cis/trans* isomerization

can deviate strongly from the solution equilibrium constant in the absence of the enzyme (Fig. 3). For Cyp18 in the presence of a tetrapeptide substrate, K_{int} approached a value near unity thus rendering the rotational barrier to show a more symmetric shape. Burbaum et al. suggested that for a catalytically optimized enzyme, K_{int} depends on how close to equilibrium the enzyme can maintain its reaction under physiological conditions (Burbaum et al. 1989). According to this model, the magnitude of the internal equilibrium constant for Cyp18 thus points to a physiological function of cyclophilins for the catalytic interconversion of native-state isomers under reversible conditions. Other peptide bond *cis/trans* isomerases have not yet been evaluated for this parameter.

However, it is unlikely that nature succeeded in implementing four different physiological functions based on a very small polypeptide scaffold, as indicated at the top of this chapter. The possibility should, of course, be kept in mind that a single mechanism could account for the prominent cellular function using the other mechanisms as functional auxiliaries.

Among the four mechanisms referred to above for the enzymes, the possibility of an exclusive role of proline-directed binding function (Schreiber and Crabtree 1992; Luban 1996) seems remote. Comparative investigations of ground state and transition state affinities of diastereomeric substrate pairs (Schiene et al. 1998) showed that chiral discrimination has evolved based on catalysis but not binding. Site-directed mutagenesis experiments on FKBPs have also added significant information about the importance of catalytic function (Fischer et al. 1998). The other biochemical functions deserve a more detailed inspection, which will be done in the sections entitled "Do peptide bond *cis/trans* isomerases catalyze folding or just play a role in holding polypeptides?" and "Peptide bond *cis/trans* isomerases as presenter proteins in immunosuppression."

Monitoring peptide bond *cis/trans* isomerase activity

Owing to their significance, the peptide bond *cis/trans* isomerization resulted in the invention of many assays. There is an intuitive feeling that assays monitoring the dynamics of peptide bond *cis/trans* isomerization are suitable for enzymatic studies as well. Oligopeptides frequently serve the role of assay substrate in vitro because the number of peptide bonds, the potential substrate sites, is not too large. The limited data that are available for the conformation that relate to a particular peptide bond in a polypeptide substrate signalize considerable difficulties for natural substrates to be assayed within cells. In fact, signal limitations occur that relate to low substrate concentrations, the reversible type of reaction, a multitude of potential substrate sites, competing polypeptide chains, and the lack of persistent chain modifications when the reaction product is formed. These limitations match partially, but not fully, the characteristics of oligopeptide substrate-based assays in vitro. Table 1 gives a summary of the assay methods available at present. These can be subdivided into assays that make use of an altered spectroscopic signal following rapid equilibrium perturbation, and those that measure directly the dynamics of conformational equilibria, preferably by NMR methods.

The isomer-specific proteolysis of the *cis/trans* isomer mixture of Suc-Xaa-Yaa-Pro-Zaa-4-nitroanilide (Suc, succinyl; Xaa, Yaa, Zaa, natural and unnatural amino acids) is presently the most popular probe for PPIase activity. It is a kinetic probe, because a slow kinetic phase follows the rapid release of 4-nitroaniline from the *trans* isomer that gives rise to an accelerated kinetic phase in the presence of a PPIase. Coupling between reversible isomerization and irreversible proteolysis let us determine rate constants for the *cis* to *trans* interconversion exclusively (Fischer et al. 1984a). Based on this principle, a semiautomated 96-well plate assay is available for screening of compound libraries (Küllertz et al. 1998). There is a correlation between the *cis* content in aqueous solution and the reliability of the assay: the larger the *cis* fraction, the better the signal-noise ratio of the assay. Dissolving a substrate containing the -Ala-Pro- sequence in 0.5 M LiCl/anhydrous TFE improves the signal/noise ratio of the assay because of an increased fraction of the *cis* isomer (Kofron et al. 1991). This protease-coupled assay was adopted to measure substrates varying in sequence, using many isomer-specific proteases like prolyl endopeptidase and dipeptidyl peptidase IV (Fischer et al. 1983), subtilisin (Schutkowski et al. 1996a), chymotrypsin (Fischer et al. 1984a), trypsin (Lin and Brandts 1985), angiotensin-converting enzyme (Merker et al. 1993), HIV-1 protease (Vance et al. 1997a) or pepsin (Vance et al. 1997b), and can be exploited to compare the activity of different PPIases using a single type of substrate. Replacing the 4-nitroanilide residue by other chromogenic or fluorogenic moieties renders the assay broadly applicable.

While the protease-coupled assay is easy to perform, side reactions such as uncontrolled proteolysis of polypeptide substrates and proteolytic degradation of numerous peptide bond *cis/trans* isomerases seriously limit the potential of this assay.

Using similar peptide substrates, many protease-free assays have been developed. The assay principle is based on the reestablishment of an isomeric equilibrium after jumping in native buffer conditions from nonaqueous solvents and extreme pH values (Garcia-Echeverria et al. 1992; Wüthrich and Grathwohl 1974; Schiene-Fischer and Fischer 2001; Zhang et al. 2002). Interestingly, considerable perturbation for the ratio of the isomeric fractions of a prolyl bond occur when oscillating about the critical micelle concentration (Kramer and Fischer 1997).

Assuming isomer-specific differences of UV/Vis spectra, NMR signals, fluorescence quenching, or chemical properties (pK_a values), the time-courses of re-equilibration provide a signal that is determined by the sum of the first-order rate constants for the *cis* to *trans* and *trans* to *cis* interconversion (Janowski et al. 1997).

Given the experimental fact that amplitude of the signals arising from solvent perturbations is usually very small, *cis/trans* ratio enhancement achieved by covalent modifications of the substrate peptides allowed improvement in the signal quality of the assay. By intramolecular disulfide bridging, the whole population of 2-aminobenzoyl-Cys-Phe-Pro-Ala-Cys-4-nitroanilide experienced a *cis* Phe-Pro moiety that declined after disulfide ring reduction to an open chain *cis* isomer level of about 15% (Weisshoff et al. 1995). In the presence of high concentrations of reductive agent, as dithiothreitol or tris-(2-car-

Table 1 Assays for peptide bond *cis/trans* isomerases

Method	Detection	Preferred reaction direction	Substrate type	Specificity	Reference
Isomer specific proteolysis (UV/Vis)	Proteolysis, release of 4-nitroaniline, 390 nm	cis→trans	Oligopeptides	Prolyl bond	Lin and Brandts 1979; Fischer et al. 1984a; Fischer et al. 1984b; Lin and Brandts 1985; Garcia-Echeverria et al. 1992; Küllertz et al. 1998
Capillary zone electrophoresis (UV/Vis)	Proteolysis, peptide bond 200 nm	cis→trans	Thioxo-oligopeptides	Thioxo-prolyl bond	Meyer et al. 1994
LiCl/TFE solvent jump (UV/Vis)	Tyr 3-NO_2 428 nm	cis⇌trans	Oligopeptides	Prolyl bond	Garcia-Echeverria et al. 1993
	Phe 4-NO_2 330 nm; Phe 2,4-F_2 246 nm	cis⇌trans	Oligopeptides	Prolyl bond	Janowski et al. 1997
	Peptide bond 220 nm	cis⇌trans	Oligopeptides	Prolyl bond	Zhang et al. 2002
	Many ^{1}H-NMR-signals	cis⇌trans	Oligopeptides	Prolyl bond	Reimer et al. 1998
LiCl/TFE solvent jump (fluorescence)	Aminobenzoic acid Ex. 337 nm, Em. 410 nm	cis⇌trans	Oligopeptides	Prolyl bond	Garcia-Echeverria et al. 1992
Caged conformation (fluorescence)	Aminobenzoic acid Ex. 320 nm, Em. 410 nm	cis⇌trans	Cyclic disulfide peptides	Prolyl bond	S. Fuessel, unpublished data
pH jump	Peptide bond 220 nm	cis⇌trans	Dipeptides/ tripeptides	Prolyl/ nonproyl bond	Schiene-Fischer and Fischer 2001 Schiene-Fischer et al. 2002a Zhang et al. 2002
	^{1}H-NMR	cis⇌trans	Tripeptides	Prolyl bond	Hsu et al. 1990
Micelles, jump below CMC	Proteolysis, release of 4-nitroaniline, 390 nm	cis⇌trans	Oligopeptides	Prolyl bond	Kramer and Fischer 1997
1D-^{1}H-NMR	Line shape	cis⇌trans	Oligopeptides	Prolyl/nonprolyl bond	Hübner et al. 1991; Kern et al. 1993; Kern et al. 1995
	Saturation transfer	cis⇌trans	Oligopeptides/ short proteins	Prolyl/nonprolyl bond	Evans et al. 1989; Hsu et al. 1990
2D-^{1}H-NOESY-NMR	Magnetization transfer	cis⇌trans	Oligopeptides/ short proteins	Prolyl/nonprolyl bond	Evans et al. 1989; Kern et al. 1993 Scherer et al. 1998; Schiene-Fischer et al. 2002a

Table 1 (continued)

Method	Detection	Preferred reaction direction	Substrate type	Specificity	Reference
1D-^{19}F-NMR	Magnetization transfer ^{19}F	cis⇌trans	Oligopeptides	Prolyl/nonprolyl bond	London et al. 1990
Refolding/unfolding	Intrinsic fluorescence	trans→cis cis→trans	Proteins	Prolyl/nonprolyl bond	Lang et al. 1987; Mücke and Schmid 1992; Schiene-Fischer et al. 2002a
	UV/Vis	trans→cis cis→trans	Proteins	Prolyl bond	Lang et al. 1987
	Circular dicroism	trans→cis cis→trans	Proteins	Prolyl bond	Bächinger 1987; Thomson et al. 1989
	Proteolysis	trans→cis cis→trans	Proteins	Prolyl bond	Bächinger 1987; Davis et al. 1989
	Enzymatic activity	trans→cis cis→trans	Proteins	Prolyl bond	Lin et al. 1988; Kiefhaber et al. 1990a; Chiti et al. 1999

boxyethyl)-phosphine, ring opening is a rapid reaction followed by the slow equilibration of the Phe-Pro moiety. Efficient fluorescence quenching of the 2-aminobenzoyl residue by the 4-nitroanilide moiety requires the *cis* conformation. Thus, subsequent to the reductive pulse, the time-course of the fluorescence increase corresponds to the *cis/trans* interconversion in the open chain peptide that will reveal the presence of a PPIase (Fischer et al. 2001). Usually, most substrates and PPIases are not susceptible to interference by the reductive agents.

Capillary zone electrophoresis (CZE) has the potential of serving as a probe for PPIase activity because separation of conformational isomers is possible at low temperatures, as was already shown for the calcium regulatory hormone calcitonin. (Thunecke and Fischer 1998).

Furthermore, the potential to resolve the isomeric state of prolyl bonds in proteins was recently shown by peak splitting in ion mobility/time-of-flight mass spectroscopy (Counterman and Clemmer 2002).

In combination with isomer-specific chemical shift dispersion, dynamic NMR spectroscopy provides a powerful tool for locating isomerizing peptide bonds in polypeptides, and for determining rate constants, but is limited to high substrate concentrations. It generally does not require equilibrium perturbation by chemical means but instead transferred magnetization to isomers.

One-dimensional ^{1}H-NMR experiments revealed slow dynamics in prolyl bond isomerization; line broadening is not significant for the *cis* and *trans* signals. Measurements in the presence of peptide bond *cis/trans* isomerases show line broadening and coalescence (Hübner et al. 1991) that can be quantified by comparison with simulated line shapes based on quantum mechanical density matrix formalism (Binsch 1969; Kern et al. 1995; Videen et al. 1994).

2D-NOESY experiments are more sensitive to catalytically enhanced interconversion rates than 1D NMR techniques. Exchange cross-peak intensity at different mixing times gives access to the rate constants of interconversion (Baine 1986).

The cross-peak evaluation is associated with the information about enzyme specificity as exemplified by human calcitonin. Cyp18 is able to catalyze only one of the two prolyl bonds present in the molecule (Kern et al. 1993). This technique allows for monitoring prolyl bond isomerizations in medium-sized proteins (Mallis et al. 2002; Bosco et al. 2002).

Since the same Bloch equations govern the magnetic exchange during the mixing time of a the 2D-NOESY experiments and the perturbation transfer time, saturation transfer studies provide another possibility for gaining information about conformational processes. This is done by irradiating a specific small frequency range of the ^{1}H spectrum containing one of the conformers under investigation. In the case of conformational dynamics, the recorded spectrum shows peaks at the positions assignable to the different species in equilibrium (Hsu et al. 1990; McCornack et al. 1997). For instance, investigations of staphylococcal nuclease at elevated temperature close to the denaturation transition point by NMR revealed four states that were assigned to *cis* and *trans* conformers of the Pro117 prolyl bond in the native and unfolded state, respectively (Evans et al. 1989).

There is no difference in principle between the situation of prolyl bonds and secondary amide peptide bonds with respect to isomer-specific chemical shift dispersion, and NMR sensitivity of interconversion dynamics. In fact, in cases in which the isomer-specific dispersion is large, the small size of the *cis* population does not hamper the determination of cross-peak intensities (Scherer et al. 1998). Recently, this technique has been utilized to detect DnaK catalysis of the Ala-Tyr peptide bond in Ala-Ala-Tyr-Ala-Ala, and to rule out the possibility of a DnaK-catalyzed Tyr-Ala interconversion (Schiene-Fischer et al. 2002a).

Not only ^{1}H nuclei but also ^{19}F nuclei in labeled amino acids has been used as a spectroscopic probe as shown for a Cyp18-catalyzed isomerization reaction in a derivative of the tissue hormone bradykinin (London et al. 1990).

Refolding of denatured proteins has often been used to investigate PPIase effects on folding kinetics. These time-courses may in turn serve as a tool for assaying PPIase activity.

When chemically denatured, both wild-type ribonuclease T1 (wtRNaseT1) and various site-directed mutagenized protein variants turned out to be suitable substrates for PPIases and APIases.

Two of the prolyl bonds of RNase T1 (Tyr38-Pro39, Ser54-Pro55) adopt the *cis* conformation in the native state, but have a marked reduction in the *cis* population in the unfolded state. Thus, the *trans* to *cis* isomerization is a quasi-irreversible process because the reverse isomerization is disfavored under strongly native refolding conditions. The refolding kinetics can be probed by fluorescence-, CD-, UV/Vis-, real-time NMR-spectroscopy, and the regain of enzymatic activity (Thomson et al. 1989; Kiefhaber et al. 1990a; Balbach et al. 1999). Recently, time-resolved FTIR-difference-spectroscopy has also been applied successfully (Moritz et al. 2002). A complete kinetic folding model has been established by unfolding-refolding double mixing experiments (Kiefhaber et al. 1990b; Mayr and Schmid 1993; Mayr et al. 1996). If RNase T1 is denatured, reduced, and the cysteine side chains carboxmethylated it is still able to adopt an active, native-like conformation at high ionic strength. Thus, instead of applying denaturating agents or extreme thermal conditions, variable concentrations of NaCl can be used to switch between native and unfolding conditions. In general, the PPIases tolerate this condition without interference with their enzyme activity. Using the remarkable features of RNaseT1, this enzyme has been utilized as a standard assay for PPIase catalysis on proteinaceous substrates in the refolding and the unfolding direction (Mücke and Schmid 1992). Replacement of the Pro39 for an Ala by site directed mutagenesis resulted in a protein variant of decreased stability. As verified by NMR spectroscopy, the *cis* conformation is conserved at this position resulting in a *cis* Tyr38-Ala39 bond in the native protein that is nearly completely converted to the *trans* form in the unfolded state (Mayr and Schmid 1993; Mayr et al. 1994). This RNase T1 variant allows for direct measurement of APIase activity on a native-like protein (Schiene-Fischer et al. 2002a).

Repertoire of peptide bond *cis/trans* isomerases in human cells

The classification of enzymes according to the type of the reaction they catalyze was the underlying principle of the nomenclature of peptide bond *cis/trans* isomerases. Major determinants were identified as the primary substrate specificity involving the chemical type of the reactive peptide bond and the amino acids flanking the reactive bond C-terminally. Presently, the peptide bond *cis/trans* isomerases can be classified into two groups: the APIases, and the PPIases (E.C.5.1.8; Fig. 4). This naming is based on the targeted sub-

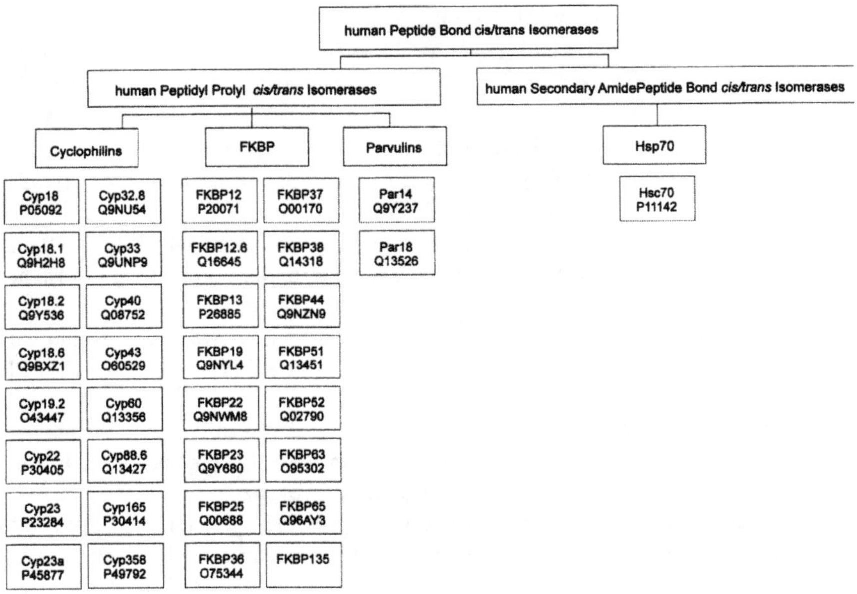

Fig. 4 Presently known human peptide bond *cis/trans* isomerases classified according to specificity- and sequence-based similarities. The individual enzymes have been identified by the SWISS-PROT/TREMBL accession number

strate bond that is unambiguously the prolyl bond for the imidic peptide bond *cis/trans* isomerases because proline represents the only canonical imino acid.

The question of whether APIases also exhibit amino acid restrictions for substrate chains as relevant for naming is not yet fully concluded for *Escherichia coli* hsp70 DnaK, and completely open for other putative members of the APIase group. As stated above, amino acid sequence similarities of the respective catalytic domain served as a criterion to assign PPIases into the different families: the cyclophilins, FK506 binding proteins, and parvulins. Amino acid sequence similarities between DnaK and archetypical PPIases are not evident. Enzymatic activity has already been shown for the majority of the human peptide bond *cis/trans* isomerases depicted in Fig. 4.

The term immunophilin has collectively been used for all Cyp und FKBP, especially in the reporting of physiological studies. The designation refers to the formation of Cyp18/cyclosporin A (CsA) and FKBP12/FK506 complexes and, in turn, the immunosuppressive character of these complexes. Because the involvement in immunosuppression of drug complexes formed by the wide diversity of the other cyclophilins or FKBP (Fig. 4) is still a matter of debate, this term does not hold anymore and should be abandoned. It also suffers from the difficulty of not providing an accurate designation of Cyp and FKBP isoenzymes (Fischer 1994a). Parvulins do not belong to immunophilins at all.

The designation of FKBP and cyclophilin isoenzymes as culled from the literature is confusing. For example, the progenitor FKBP was originally named human FKBP12, with the suffix designating the approximate molecular mass of the gene-encoded protein. In the human gene terminology, the gene was designated as FKBP1a. The human gene FKBP14 (alternative gene name FKBP22), on the other hand, has the entry name in the Swiss-Prot database FK14_human, which corresponds to the protein FKBP14 but has the designation

FKBP24 according to the rules applied to FKBP12 naming. Illogically, the same protein is referred to as FKBP22 (Swiss-Prot Q9NWM8) as well.

It is now proposed that for clarity's sake the enzyme family abbreviations be complemented by a suffix for the approximate molecular mass of the gene encoded sequence, and the organism resulting in human Cyp18 be the distinctive name for the prototypic cyclophilin (MW of 18.013 Da) from human cells. Rare cases of different enzymes with nearly identical molecular masses could be differentiated by adding a small character to the suffix (Cyp23a in Fig. 5). Occasionally, the human Cyp18, Cyp23, and Cyp22 enzymes are termed CypA, CypB, and CypD, respectively.

Family members of the ubiquitous enzyme class are found in abundance in virtually all organisms and subcellular compartments. Their amino acid sequences are highly conserved phylogenetically. Organisms express many members of each family that encompass one or more PPIase domains complemented with other functional polypeptide segments. These segments or domains have been found N-terminally and C-terminally to the catalytic domain. Figures 5, 6, and 7 illustrate the domain-like organization for peptide bond *cis/trans* isomerases presently known to exist in human. Based on three-dimensional structures, PPIases of the FKBP- and parvulin-type share a common pattern of structural motifs culminating in the existence of the FKBP-like superfold (Fig. 8; Sekerina et al. 2000; Ranganathan et al. 1997).

Do peptide bond *cis/trans* isomerases catalyze folding or just play a role in holding polypeptides?

In vitro experiments frequently reveal that peptide bond *cis/trans* isomerases encompassing more domains than just the catalytic module are able to prevent chain aggregation in typical chaperone assays. Furthermore, some PPIases were found to be uncoupled in their isomerase- and aggregation-preventing activity, and are occasionally insensitive toward small-molecule PPIase inhibitors in promoting renaturation and in suppressing protein aggregation in vitro (Furutani et al. 2000; Ou et al. 2000; Kurek et al. 2002; Bose et al. 1996; Freeman et al. 1996. Ramm and Plückthun 2000; Ramm and Plückthun 2001; Arie et al. 2001). It has been argued that the affinity of active PPIases and enzymatically deficient PPIase variants for unfolded polypeptides, which are in this respect holding chaperones, may be evidence that the catalyzed *cis/trans* isomerization is auxiliary to their chaperone function in vivo. In addition, experiments have been carried out in negative mutants of multidomain PPIase genes that investigated whether the catalytic core is dispensable in rescuing physiological defects (Behrens et al. 2001; Ramm and Plückthun 2000; Webel et al. 2000; Arie et al. 2001; Li et al. 2001b; Webb et al. 2001). Functional redundancy between enzyme class members and limitations in phenotypic analyses make the response to truncated PPIases difficult to interpret. However, the above-mentioned inhibition experiments are not suggestive of a chaperone function that evolved independently of the PPIase function. Due to the extended substrate-binding region, blocking the narrow active site region will not necessarily lead to abrogation of a large fraction of secondary binding sites. In an evolving account of the enzyme's specificity, nature frequently utilized an extended array of secondary binding sites (or catalytic subsites) in an enzyme-polypeptide substrate complex. Although secondary binding sites required for specificity enhancement release binding energy in the transition state, the flexibility of the enzyme site cavity allows prior ground state binding to occur. The nature of the reaction catalyzed by *cis/trans* isomerases

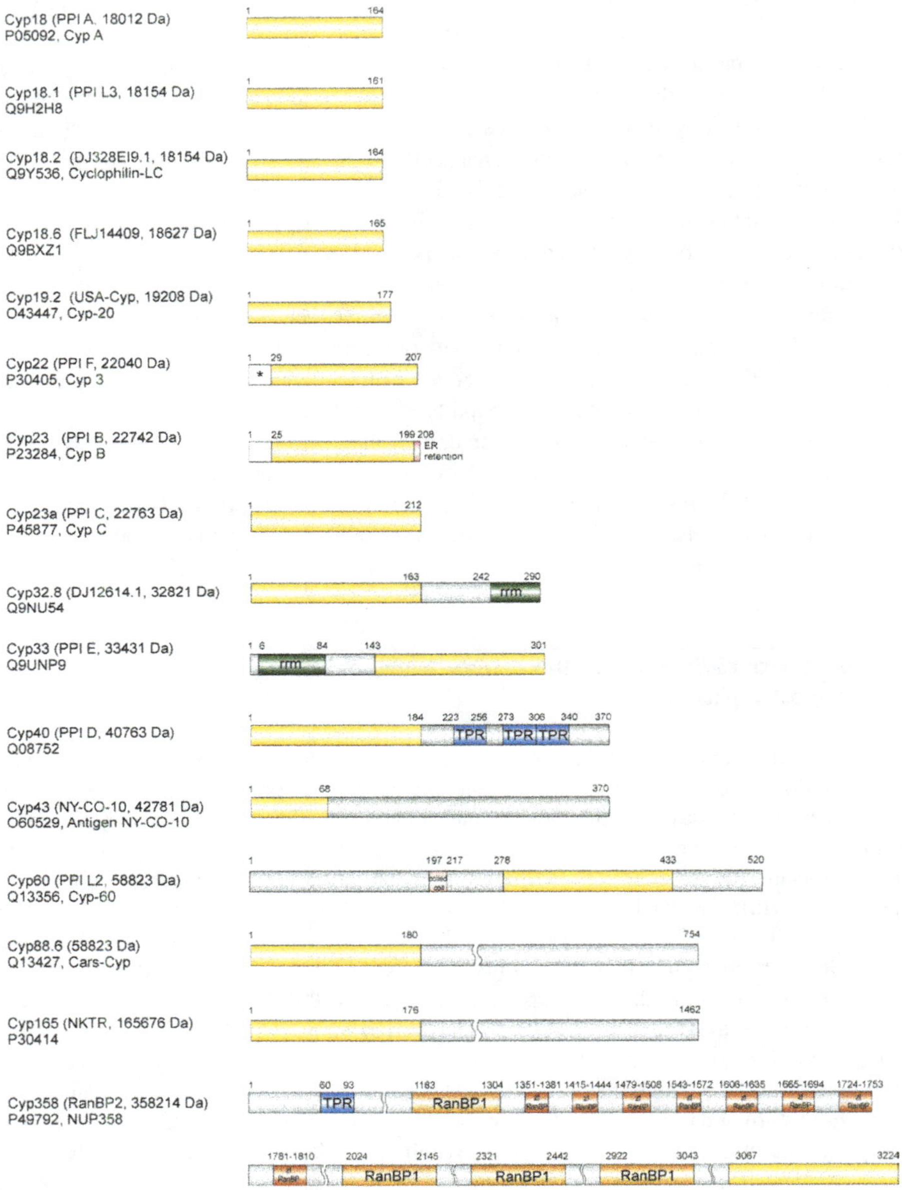

Fig. 5 Human cyclophilins. Protein nomenclature is according to Fischer (1994a). The gene name and the size of the unprocessed protein are shown in the *brackets*. In the *second row*, the accession number of SWISS-PROT/TREMBL database and one example of an alternative protein name is given. The amino acid residues that border the protein domains or modules are designated according to SWISS-PROT or Pfam databases. The cyclophilin domain is depicted in *yellow*. Signal sequence regions are shown as *colorless boxes*. *rrm* RNA recognition motif, *TPR* tetratricopeptide repeat, *RanBP1* Ran binding protein 1 domain, *zf RanBP* Zn-finger, Ran-binding; * indicates "potential"

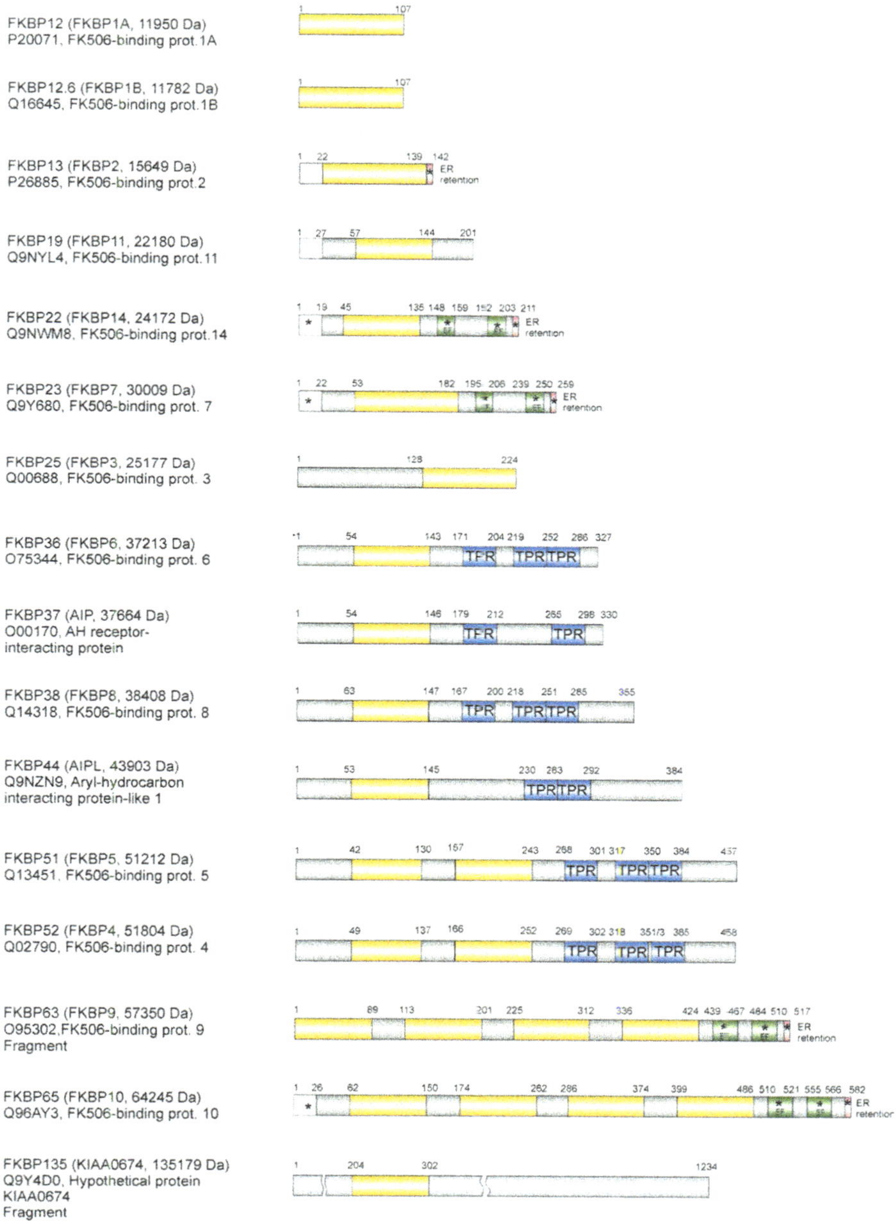

Fig. 6 Human FKBPs. Nomenclature is according to Fischer (1994a). The gene name and the size of the unprocessed protein are shown in the *brackets*. In the *second row*, the accession number of SWISS-PROT/TREMBL database and one example of an alternative protein name is given. The amino acid residues that border the protein domains or modules are designated according to SWISS-PROT or Pfam databases. The FKBP domain is depicted in *yellow*. Signal sequence regions are shown as *colorless boxes*. *EF* EF hand, *TPR* tetratricopeptide repeat; * indicates "potential"

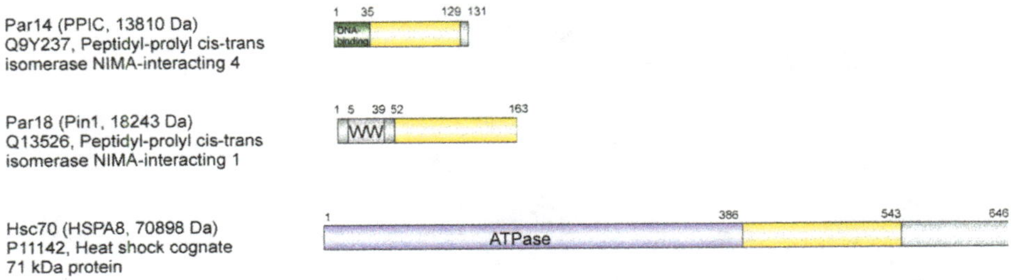

Fig. 7 Human parvulins and hsc70, a member of the hsp70 family. Nomenclature is according to Fischer (1994a). The gene name and the size of the unprocessed protein are shown in the *brackets*. In the *second row*, the accession number of SWISS-PROT/TREMBL database and one example of an alternative protein name is given. The amino acid residues that border the protein domains or modules are designated according to SWISS-PROT or Pfam databases. The parvulin and the APIase domain of hsc70 are depicted in *yellow*. *WW* WW domain. DNA binding of Par14 is according to Sekerina et al. (from Sekerina et al 2000)

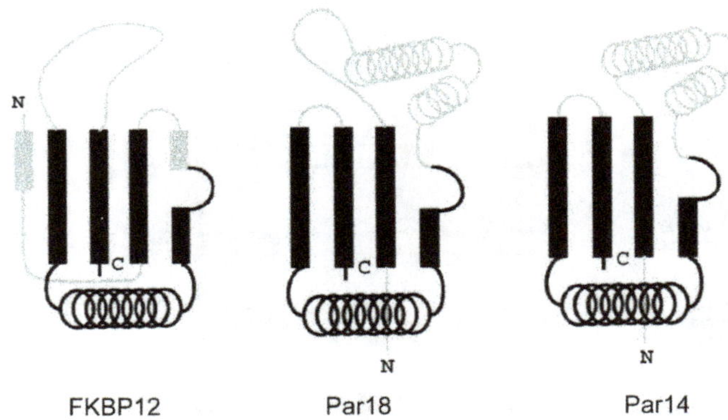

Fig. 8 The FKBP superfold. Topological arrangement of secondary structure elements of human FKBP12, human Par18 (Pin1) and human Par14. A minimal conserved structural fold is indicated by *black colored secondary structure symbols* (Sekerina et al. 2000)

requires that the fraction of enzyme-bound peptide chain is partitioned into the Michaelis complexes of the *cis* and *trans* isomer. The complexes exist in a highly mobile equilibrium, shown to be of comparable stability for Cyp18 and having very similar chemical properties (Fig. 3). Subsite-dependence of binding affinities is differently controlled for ground state and transition state interactions. Compared with a *cis* dipeptide, ground state binding to Cyp18 of a *cis* tetrapeptide is 70-fold stronger while transition state affinity increases more than 10^6 fold in this series of compounds (Fischer et al. 1994b). Thus, experiments primarily revealing ground state affinity cannot differentiate polypeptide chains underlying catalytic interconversion from passively chaperoned chains.

As stated above, enhanced polypeptide sequestration by utilizing subsite binding is secondary to the catalytic function of peptide bond *cis/trans* isomerases. Consequently, catalytic specificity has primarily evolved from the interactions of secondary binding sites. Regarding power of catalysis, the *cis/trans* isomerase function of the PPIases and the APIases is evolved perfectly, realizing relaxation times of the catalyzed *cis/trans* interconversion of up to a range of submilliseconds (Janowski et al. 1997; Kern et al. 1995; Kofron et al. 1991; Scholz et al. 1997; Schiene-Fischer et al. 2002a).

In contrast, with the exception of the ATP-dependent APIases, the peptide bond *cis/trans* isomerases do not show highly evolved chaperone properties. It can be characterized by a holding but to a much smaller extent by a refolding function. Remarkably, an archetypical PPIase cannot prevent aggregation of unfolded polypeptides consistent with the view that the additional domains complementing the PPIase core are the major determinants for high affinity ground state binding of substrates (Kern et al. 1994; Scholz et al. 1997). Despite this fact, Cyp18 presents an extended binding cleft to a 25-mer proline-containing oligopeptide from the HIV-1 gag capsid protein (residues 86–96 of gag) encompassing nine residues of the peptide substrate in direct contact with the enzyme (Zhao et al. 1997). A 151-residue fragment of capsid displays a similar extension of the binding region (Gamble et al. 1996). As a nearly perfectly evolved enzyme, however, Cyp18 exhibits relatively weak ground state affinity to substrates (Schutkowski et al. 1996b; Yoo et al. 1997). Larger PPIases might present a polypeptide-binding crevice running through several domains, including the active site cleft (Patzelt et al. 2001). A candidate polypeptide-binding site of the parvulin-like SurA extends 15 Å in length encompassing one PPIase domain and the N- and C-terminal domains (Bitto and McKay 2002).

The chain segment of a substrate interacting with Cyp18 has to adopt an extended conformation in the complex. Interestingly, comparative binding experiments revealed subtle differences in the extended chain conformation of substrates from different sources for both the FKBP-like trigger factor and the APIase DnaK (Schaffitzel et al. 2001). Usually, complexation by substrates or inhibitors did not impede marked structural changes on prototypic PPIases.

It is the ground state binding affinity underlying the interactions with extended subsites that is intrinsic to the formation of the two Michaelis complexes, the (*cis**enzyme) and (*trans**enzyme), which might possibly lead to ambiguity and confusion relating to a putative chaperone activity of peptide bond *cis/trans* isomerases in vivo. To what extent the peptide bond *cis/trans* isomerases make use of this holding function bypassing the isomerizing site in physiological systems has not been clarified. In an extreme position, substrate sequestration might be considered as the precondition for efficient catalysis with almost no additional role in the cell.

In several cases, the ground state polypeptide binding capacity of peptide bond *cis/trans* isomerases is highly specific, thus allowing functional dissection to be obtained. For example, the WW domain of the human PPIase Pin1 (Fig. 7) exhibits affinity for phosphoserine- or phosphothreonine-containing protein segments but fails to engage unphosphorylated peptides (Lu et al. 1999). When compared with the catalytically competent PPIase domain, phosphopeptide binding is much stronger for the WW domain. However, essential biological functions of Pin1 are carried out by the PPIase domain but not the WW domain, rendering it a bystander site (Lu et al. 2002). Similar to Pin1, expression of isolated fragments of FKBP52 succeeded in differentiation of the PPIase site and the chaperoning site, indicating the C-terminal tetratricopeptide repeats as the chaperoning site (Pirkl et al. 2001). In another example, the preference for short stretches of hydrophobic amino acids with no positional preferences was shown for trigger factor by a cellulose-bound peptide library (Patzelt et al. 2001). Again, the catalytic domain in isolation retains the full-length trigger factor binding specificity but at reduced affinity. It is likely that the polypeptide substrate binding sites of trigger factor extend beyond the PPIase domain. Ground state affinity for polypeptide stretches can be realized by remote binding sites of full-length trigger factor that are independent of the presence of prolyl residues (Scholz et al. 1998a; Patzelt et al. 2001). Enforced prolyl bond propinquity coupled with optimal reactive geometry for both the substrates and the active site residues with which they interact is liberated upon transition state formation. Potentially, properly located secondary amide peptide bonds of an extended substrate chain might be prone to trigger factor catalysis. However, for extended stretches of proline-free polypeptides, the kinetics of *cis/trans* isomerization is difficult to detect at a particular bond (Scholz et al. 1998b).

Interestingly, experiments performed with the carboxymethylated forms of α-lactalbumin and RNase T1 in vitro shows that the binding affinity of trigger factor for unfolded polypeptide chains is relatively low. However, the lifetime of the complexes is in the range of 100–200 ms for physiological concentrations of the trigger factor, allowing rapid scanning of the substrate chain for bonds where catalysis is to occur (Maier et al. 2001). Apparently, the ability of the multidomain PPIases to suppress aggregation and to promote renaturation is a property independent of cochaperones and ATP. Similarly, the APIases function of DnaK does not require the full ATP/DnaJ/GrpE mixture for repeated catalytic cycles of *cis/trans* isomerization in short oligopeptide substrates, and short nonnative segments in native-like folding intermediates. These types of small substrates realize only a few binding contacts when the complexes is formed on the catalytic pathway (Schiene-Fischer at al. 2002a).

In contrast, association of unfolded polypeptides with DnaK can result in tight Michaelis complexes (dissociation constants <<1 μM) where the reaction product cannot be easily released (de Crouy-Chanel et al. 1999). Under steady-state conditions, polypeptide chain liberation requires the cooperation of two accessory proteins, DnaJ and GrpE, and the hydrolysis of ATP in refolding assays (Langer et al. 1992).

To differentiate peptide bond *cis/trans* isomerase functions by analogy, proteases need to be considered. They are related enzymes in terms of the correlation between the principal enzyme properties; a particular peptide bond is targeted to chemical attack, and catalysis is accompanied by polypeptide substrate sequestration. Like peptide bond *cis/trans* isomerases, most proteases belong to enzymes that possess a number of substrate/enzyme interactions over an extended region of the protein. However, the large chemical difference between a substrate and a product of proteolysis has been used as an easy means to differentiate the substrate holding from the catalytic function. As expected, anhydrochymotrypsin, a protein variant devoid of proteolytic activity by having a modified active site serine residue, retained specific affinity for numerous hydrophobic peptides (Ohta et al. 1993). Similarly, equilibrium dissociation constants for the interaction of trypsin and anhydrotrypsin with proteinase inhibitors and short oligopeptides showed that a functional proteolytic site is not required to achieve high affinity binding (Olson et al. 1995; Carter and Wells 1987). Notably, affinity labels for the active serine side chain do not prevent polypeptide sequestration to the inactivated α-chymotrypsin (Enghild et al. 1994).

More specialized proteases might exemplify the notion that a natural function is possible for both proteolysis and chaperoning the folding of polypeptides. A proteolytically inactive mutant of the ATP-dependent *E. coli* protease Lon acts as a holding chaperone in vivo. It can complement to some extent a *lon* deletion mutant strain in UV sensitivity and capsule formation (Van Melderen and Gottesman 1999). Substrate protection from external proteolysis by sequestration was suggested but did not call the primary function of Lon as a protease into question.

Whether the respective substrates require an escort by either an enzymatic or a chaperone function might be a matter of environmental conditions. The conserved heat-shock protein DegP can exhibit both chaperone and proteolytic functions. The chaperone function dominates at low temperatures, while the proteolytic activity is present at elevated temperatures (Spiess et al. 1999). Consequently, the PPIase SurA and the protease DegP share functional redundancy under certain growth conditions, probably based on their respective periplasmic chaperone functions (Rizzitello et al. 2001).

In conclusion, polypeptide sequestration of most proteases have evolved to achieve substrate specificity, usually not playing an independent role. Similarly, the cumulative evidence indicates that holding a polypeptide chain for the sake of offering environmental protection serves a minor role in the physiological function of peptide bond *cis/trans* isomerases.

Mode of action of peptide bond *cis/trans* isomerases in cell signaling

General considerations

Although much can be learned about physiological peptide bond *cis/trans* isomerase function from ex vivo studies, it is still difficult to evaluate catalyzed *cis/trans* isomerizations in cellular signaling. Analyses were most successfully accomplished by the generation of peptide bond *cis/trans* isomerase variants attenuated in their catalytic activity and their structure-activity relationships when tested in signal cascades.

In addition, small cell-permeable effectors represent in principle powerful tools for identification of links between catalyzed peptide bond *cis/trans* isomerization and signaling processes. Most attempts to unravel physiological functions of peptide bond *cis/trans* isomerases in vivo have been done by this pharmacological approach. In particular, the immunosuppressants cyclosporin A (CsA), FK506, and rapamycin have been proposed as biochemical markers and drugs for therapeutic intervention that may provide results useful for dissecting functions in vivo. These compounds are competitive inhibitors of almost all cyclophilins or FKBPs present in human tissues, but to a different extent.

However, as already discussed in the previous sections, a drug effect on a cellular signal cannot be attributed solely to reduced PPIase activity. The inhibitors might also affect proline-directed PPIase/ligand associations, may liberate gain-of-function effects, and interfere with the sequestration of polypeptides.

In conjunction with the great number of isoenzymes potentially affected (see Fig. 4), these four mechanisms, potentially active either in isolation or in parallel, impose severe limitations on the interpretation of the biological effects obtained by application of these drugs.

Fortunately, there are several choices of CsA and FK506 derivatives that allow us to select an inhibition regimen that is more useful for the discrimination between the individual mechanisms of action, allowing for more reliable predictions of the physiological function of peptide bond *cis/trans* isomerases. Some derivatives are high-affinity inhibitors, like [*N*-Methylalanine]^6-cyclosporin A (Sigal et al 1991) and 506BD (Somers et al 1991), toward their respective binding proteins but devoid of the immunosuppressive properties seen in parent drugs. Often these compounds are termed nonimmunosuppressive (or monofunctional) derivatives because of their inability to activate the presenter protein function of their respective prototypic binding proteins.

Even in this favorable circumstance the fact that inhibitors are not yet available that can differentiate among the paralogs of Cyp18 or FKBP12 imposes limitations on the interpretation of the effects of nonimmunosuppressive derivatives in in vivo studies. It must also be noted that there is no reversible, nonpeptidic inhibitor published for the APIase and the PPIase activity of DnaK and the parvulins, respectively.

However, data on the degradation of the proline-containing peptide hormone bradykinin present a clear example of the reliable conclusions that can be drawn from drug application, provided the in vivo experiment informs us about both the substrate and the enzyme state. As already discussed in the section "Monitoring peptide bond *cis/trans* isomerase activity," the proteolytic inactivation of proline-containing oligopeptides like bradykinin is prone to prolyl bond-mediated isomer-specific effects. In the isolated perfused rat lung the limitations that *cis* prolyl bonds can apply to the degradation of peptide hormones have been investigated in the presence of cyclophilins. The PPIases can abolish the fraction of the *cis* isomer of bradykinin surviving inactivation by angiotensin-converting enzyme during a single transit through the pulmonary circulation. Application of CsA reverses the effects of cyclophilin. In this study, the result connects the mode of action of cyclophilins under physiological conditions with the mechanism of action (1) ("Reaction profile of peptide bond *cis/trans* isomerases"), bypassing the other mechanistic possibilities (Merker and Dawson 1995).

pSer(Thr)-Pro-directed peptidyl prolyl *cis/trans* isomerases

Structure-activity relationships of enzyme-variants modified by site-directed mutagenesis can also serve as a valuable tool for exploring the determinants of peptide bond *cis/trans* isomerase function.

As the Pin1-like PPIases exhibit a high level of substrate specificity, functional redundancy is less probable for these enzymes, as readily interpretable results may be obtained from experiments with Pin1-like enzyme variants attenuated in catalytic activity (Yaffe et al. 1997).

Pin1 was found to be required for the DNA replication checkpoint in *Xenopus laevis*. This defect in replication checkpoint function was reversed after the addition of recombinant *Xenopus* wtPin1, but not the Pin1 Cys109Ala variant to the wtPin1 depleted extract. This Pin1 variant has less than 0.1% residual enzyme activity while phosphopeptide-binding avidity is less compromised. Thus, the WW-domain-directed association of Pin1 with a multitude of phosphoproteins does not constitute the prime cause of the physiological effects of Pin1 (Winkler et al. 2000) but catalytic activity does. Similarly, complementation experiments with this Pin1 Cys109Ala variant failed to rescue lethal ESS1 depletion in budding yeast, whereas the wtPin1 has succeeded in rescuing.

The mechanism by which Pin1 exerted its critical role became obvious by combining in vitro studies on isomer-specific dephosphorylation by protein phosphatase 2a (PP2a) with dephosphorylation studies in *Xenopus* mitotic extracts, and rescue experiments in yeast (Zhou et al. 2000; Lu et al. 2002). For example, the cell cycle regulatory protein phosphatase Cdc25C contains the pThr48-Pro- and pThr67-Pro motifs that represent critical regulatory phosphorylation sites. For PP2a, the inability to dephosphorylate *cis* pThr(Ser)-Pro moieties in oligopeptides is retained for the native Cdc25C protein, and can be detected in vivo by a reciprocal genetic interactions in temperature-sensitive PP2a-deficient *PPH* and the Pin1-homolog-deficient *ESS1/PTF1* mutant strains of budding yeast. Transgenic expression of point-mutated variants of Pin1 and isolated Pin1 domains in yeast indicates that the enzymatically active PPIase domain is necessary and sufficient to carry out the essential function of Pin1 in the cell cycle. The WW domain alone, although exhibiting affinity to the pThr(Ser)-Pro-containing polypeptide chains, and enzymatically inactive full-length Pin1 variants are unable to support growth. Transfected HeLa cells overexpressing activity-depleted mutant Pin1 showed an increased frequency of apoptotic nuclei when compared with control cells overexpressing wild-type enzyme. Dramatic effects occur when Pin1 variants such as the PPIase-inactive Pin1Cys113Ala, which is still able to compete for essential phosphoproteins with endogenous wtPin1, were compared with other Pin1 variants. For Pin1Arg68,69Leu, both phosphoprotein binding to the active site and PPIase activity are lacking. Thus, transfected HeLa cells do not reveal dominant negative effects, and have lower mitotic indices (Rippmann et al. 2000).

Consequently, the catalytic attenuation of isomer-specific interactions of phosphoprotein substrates with protein phosphatases is found to have a singular role for the physiological function of Pin1-like enzymes.

PPIase effects on phosphate transfer reactions are not restricted to protein dephosphorylation. Proline-directed protein kinases can make use of a correct signature sequences for Thr(Ser)-Pro phosphorylation, provided, however, that the conformational state of the peptide bond following serine (threonine) is in *trans* conformation. Erk2-catalyzed phosphorylation rates were measured for proline-containing oligopeptides, and catalytic acceleration by FKBP12 or Cyp18 was evident in a certain time-window (Weiwad et al. 2000). Experiments that would bear most directly on the question of PPIase-catalyzed protein phosphorylation in cell signaling are those on the effects of externally added FKBPs or cyclophilins to cell extracts depleted of PPIase activity.

Thus, results on phosphate transfer reactions suggests that isomer-specificity is effective in conferring catalytic effects of a peptide bond *cis/trans* isomerase to cell signaling. The other mechanisms do not play a role in these posttranslational reactions. Interestingly, the observation of catalytic effects of PPIases on phosphatase-catalyzed and kinase-catalyzed phosphate transfer reactions allowed for the detection of *cis* pThr(Ser)-Pro bonds and *cis* Thr(Ser)-Pro bonds, respectively, in the native state of proteins, information that cannot be obtained by other methods under in vivo conditions.

Attempts were made based on immunoprecipitation, protein digestion, and protein phosphatase activity regulation to identify catalytic effects of Pin1 on native proteins in vitro. Comparing wtPin1 and activity-depleted Pin1variants, effects that might relate to catalyzed conformational changes were detected but the underlying structural changes have remained elusive (Stukenberg and Kirschner 2001). For example, the Pin1-assisted proteolytic fragment pattern of matrix-bound phosphorylated Cdc25c differs from that of a Pin1-free control. Obviously, subtle changes in the stability of the folded state that likely occur on matrix binding, protease action, and phosphorylation have to be taken into account. In a careful study of the nature of the cyclophilin-dependent steps in the folding of mouse dihydrofolate reductase, some of the pitfalls that must be overcome have been identified (von Ahsen et al. 2000).

Cyclophilins in cell signaling

Direct evidence for the functional role of cyclophilins has come from signals associated with oxidative stress. Familial amyotrophic lateral sclerosis (FALS)-associated mutant Cu/Zn superoxide dismutase-l (SOD) induces apoptosis of neuronal cells in culture associated with high levels of reactive oxygen species. An increased protein turnover might result from altered cellular free radical balance that requires an increased level of folding-supportive enzymes. Thus, the apoptosis-enhancing effect of cyclosporin A and FK506, as well as nonimmunosuppressant analogs of cyclosporin A, in the PC12 cells infected with the AdV-SODVal148Gly and AdV-SODAla4Val variants might result from the increased need for PPIase activity in this oxidative stress situation. The importance of enzymatic activity in mutant SOD-mediated apoptosis was supported by experiments showing that overexpressed wtCyp18, but not Cyp18 Arg55Ala variant with a reduced catalytic activity of about 0.1% of wild-type enzyme, protected cells from death after SODVal148Gly expression (Lee et al. 1999). In this study, a positive correlation existed between the action of PPIase inhibitors on cell death and effects of activity compromised PPIase variants. In general, the experiments reported here, which make use of nonimmunosuppressive PPIase inhibitors and transfected enzyme variants with reduced PPIase activity in parallel, provide the most reliable and general indicator for the role of peptide bond *cis/trans* isomerases in cell signaling available today. The involvement of Cyp18 in the activation of the caspase cascade in neuronal cells has been suggested by antisense technology. It has been concluded that neuroprotection by CsA against excitotoxin-induced apoptosis is, at least partly, due to inhibition of Cyp18 (Capano et al. 2002).

Similar experiments have been conducted in which inhibition of opening of the mitochondrial permeability transition pore by cyclophilin inhibitors antagonizes mitochondrial-mediated apoptosis. Pore opening is inhibited by compounds with the same affinity as they inhibit the PPIase activity of Cyp22 (Halestrap et al. 2002). On the other hand, Cyp22 Arg96Gly and His167Gln point mutations were found to completely abrogate enzymatic activity. The consequent loss of protective effects on latency of the mitochondrial membrane potential during oxidative stress in stably transfected HEK293 cells indicates a mechanism of action including a transient binding of a substrate polypeptide to the active site. Compared to control, overexpressing wtCyp22 prolongs membrane potential collapse time significantly. Furthermore, Cyp22 binding to the adenine nucleotide translocase does not depend on the active site accessibility of the PPIase, while pore opening does (Clarke et al. 2002). Thus, depending on its PPIase activity, the mitochondrial matrix enzyme Cyp22 exerts a protective effect by acting on a still unknown substrate that, in turn, indirectly inhibits pore opening. Therefore, Cyp22 has been considered a cell survival molecule (Lin and Lechleiter 2002).

In judging peptide bond *cis/trans* isomerase mechanisms, functional cell assays using the external delivery of these enzymes compare favorably to stable transfection of enzyme genes.

Although the amino acid sequence of Cyp18 indicates intracellular localization, protein secretion was shown under various conditions (Xu et al. 1992). Once secreted, the Cyp18 exerts various extracellular effector activities on target cells.

The signaling receptor for Cyp18 (and for its ER-directed Cyp23 homolog) was demonstrated to be the type I integral membrane glycoprotein CD147, which is expressed in a wide variety of body cells (Yurchenko et al. 2001; Yurchenko et al. 2002).

Whole cell experiments assaying the phosphorylation status of extracellular signal-regulated protein kinases in the presence of a panel of Cyp18 variants served to display possible initiation of downstream signaling events. Whereas externally added wtCyp18 fully enhances the level of ERK phosphorylation in CHO.CD147 cells but not in CHO cells lacking induced CD147 expression, Cyp18 variants are only effective according to their level of residual PPIase activity. Site-directed mutagenesis of the CD147 extracellular domain revealed that the Asp179-Pro180 segment was the Cyp18-directed substrate region. Neutrophil chemotaxis, a downstream signaling effect of CD147 stimulation by Cyp18, is also dependent on the intact Asp179-Pro180 recognition segment. However, Cyp18 variants have not yet been investigated for this process (Yurchenko et al. 2002; Bukrinsky et al. 2002). Unexpectedly, by studying CsA-mediated depletion of Cyp18-induced chemotaxis, nonimmunosuppressive Cyp18 inhibitors proved to be inactive (Xu et al. 1992; Leiva and Lyttle 1992). Direct binding of Cyp18 to CD147 has not yet been demonstrated, and has proven difficult for any productive interaction between the active site of a PPIases and good substrates. One justification for this suggestion is the fact that good enzymes are evolved for maximum rates, not binding. Available kinetic parameters reveal the peptide bond *cis/trans* isomerases are approaching perfectly evolved enzymes (Fischer et al. 1998).

Vascular smooth muscle cells secrete Cyp18 in response to reactive oxygen species (ROS). The secreted enzyme stimulates cell growth and protects cells against apoptosis. The Cyp18 Arg55Ala variant failed to stimulate ERK activation typical of early response to wtCyp18 (Jin et al. 2000). However, the involvement of CD147 in this type of signaling was not yet shown.

Whether these findings extend into Cyp18-dependent HIV-1 replication and whether the PPIase activity of host cell Cyp18 plays a crucial role in virus infectivity remains a matter of debate. We know now that HIV-1 entry depends on an interaction between virus-associated Cyp18 and CD147 on a target cell. Viruses whose replication did not require Cyp18 were resistant to the inhibitory effect of anti-CD147 antibody (Pushkarsky et al. 2001). By analogy, a crucial role of PPIase activity for all functional Cyp18/CD147 interactions must be considered. The capacity of nonimmunosuppressive Cyp18 inhibitors to interfere with the replication of HIV-1 virus in newly infected cells is in accord with this model, and rules out a presenter protein mechanism (Franke and Luban 1996; Bartz et al. 1995). At variance, a chimera of a Cyp18 variant fused to viral protein R (Vpr), Vpr-Cyp18, which bypasses the capsid/Cyp18 interaction in the wild-type Cyp18 chimera, exists. Despite the lack of PPIase activity in Vpr-Cyp18 variant, HIV-1 replication was rescued in a manner similar to the wild-type construct (Saphire et al. 2002). It is a noteworthy feature of Cyp18 that its combined properties as a secreted peptide bond *cis/trans* isomerase and an extracellular cytokine have much in common with the other subclass of peptide bond *cis/trans* isomerases, the APIases of the Hsp70 family (Asea et al. 2000).

The possible involvement of the PPIase activity in the Cyp23-induced adhesion of the CD4(+)CD45RO(+) subset of T cells to the extracellular matrix appears to be less suggestive, although CsA decreases the percentage of the T cells that adhere to fibronectin-coated wells (Allain et al. 2002). The simultaneous presence of high-affinity glycosaminoglycans and CD147 binding sites on T cells renders functional studies difficult to decode. Whereas the Cyp23 active site was not required for adherence of T cells to Cyp23 coated wells, the presence of activity-depleted Cyp23 variants in a soluble form shows a reduction in the enhancement of cellular adhesion to coated fibronectin induced by wtCyp23. A downstream signaling process was suggested that relates stringently to a catalyzed prolyl bond isomerization (Carpentier et al. 2002).

Beside its C-terminal ER retention signal, the cleavable leader sequence of Cyp23 contains a putative nuclear translocation signal that controls the retrotranslocation of the somatolactogenic hormone prolactin into the nucleus (Rycyzyn et al. 2000). The Cyp23-mediated potentation of the prolactin-driven proliferation of rat T lymphoma cell line Nb2 cells is up to 18-fold with a concurrent 4.5-fold increase in viable cell number over prolactin stimulation alone. Despite unaltered prolactin affinity of the Cyp23 Arg63Ala and Cyp23 Phe68Ala variants to prolactin, the enzymatically inactive mutant enzymes do not influence the activity of a prolactin responsive gene in various luciferase reporter gene studies. The nuclear translocation of the prolactin is independent of the PPIase activity of Cyp23. In contrast, the wtCyp23-enhanced Stat5 transcriptional activity upon prolactin administration appears to be related to a catalyzed conformational change within Stat5 that, in turn, mediates the release of the Stat5 inhibitor PIAS3. The location of the regulatory prolyl bond in Stat5 remained unknown (Rycyzyn and Clevenger 2002).

The conformational polymorphism intrinsic to proline-containing polypeptides has been linked to transcriptional control in several instances, and was suggested as playing a principal role in transcription (Shaw 2002). In temperature-sensitive ESS1 mutants of yeast, multicopy suppressors that rescue the mitotic-lethal phenotype are involved in transcription. In addition to Cyp17, other suppressors exist that bind DNA, modify chromatin structure, or are regulatory subunits of RNA polymerase II. For example, Ess1/Pin1 interacts with the C-terminal domain (CTD) of RNA polymerase II in vitro and in vivo, thereby abolishing an isomer-specific interaction with proteins required for transcription (Wu et al. 2000). In Ess1-deficient cells, Cyp17 has a compensatory potential that strictly depends on a functional PPIase site. The findings support a common target for Ess1 and cyclophilin. Both PPIases are able to modulate the activity of the Sin3-Rpd3 histone deacetylase complex, and excess histone deacetylation causes mitotic arrest in ess1 mutants (Arevalo-Rodriguez et al. 2000).

Using an electrophoretic mobility shift (EMS) assay, the c-Myb DNA binding activity in nuclear extracts of human Jurkat T cells is inactivated by Cyp40 in a CsA-sensitive manner. Cyp40 binding to and Cyp40 catalysis of c-Myb could be separated based on activity-function relationships, suggesting that a functional catalytic site is the prerequisite for the negative regulation of DNA binding. In contrast to c-Myb, v-Myb did not respond to Cyp40 in the EMS assay. Yet, unrecognized mutations in the DNA binding domain probably allowed v-Myb to escape Cyp40-mediated downregulation (Leverson and Ness 1998). Unlike Cyp40, nuclear FKBP25 increased DNA binding to the transcription factor YY1 without having any FKBP25 bound to the DNA complex. Whether enzymatically inactive FKBP25 variants are capable of increasing DNA binding is not yet known (Yang et al. 2001). A functional interaction was also shown between the IRF-4 member of the interferon regulatory factor family of transcription factors and FKBP52. Ascomycin sensitivity of the interferon-stimulated response element probe lead to the suggestion that FKBP52-inhibitor-mediated restoration of binding between IRF-4 requires an intact active site of the PPIase. Both immunoblot analysis and limited proteolysis of IRF-4 indicated a structural modification implemented by association of FKBP52 (Mamane et al. 2000; Sharma et al. 2000). The IRF-4 protein plays an important role as a regulator of human T cell leukemia virus type 1 (HTLV-1) gene transcription.

It seems difficult to explain the above in vitro effects on transcription factors based on enzyme catalysis because peptide bond *cis/trans* isomerases cannot shift a conformational equilibrium toward biologically active or inactive state when present in catalytic amounts.

Metabolic energy as a specific means to shift equilibria is also lacking in these experiments. Only if the conformational interconversion is coupled to an irreversible reaction or the concentration of the PPIase exceeds that for the substrate, do the conditions determining the equilibrium of the free conformers change, to be followed by a new equilibrium. Nonetheless, the reactivity differences are only transient, and require a typical time window of observation. This is not to seen in the EMS assay reporting Cyp40 effects, for example.

However, metastable, native-like folding intermediates are likely to play a key role in PPIase effects (Schiene-Fischer and Yu 2001). Kinetically trapped high-energy c-Myb conformers of different DNA binding activity should therefore be considered in explanations of the dramatic regulatory effects of the enzyme activity of Cyp40.

While the cyclophilins, FKBPs, and hsp70 that take part in the assembly of receptor complexes continue to grow, the involvement of an enzyme or chaperone function in natural receptor response of all but a few remain obscure. According to a recent review, many receptor complexes are under intensive investigations (Schiene-Fischer and Yu 2001).

As a receptor tyrosine kinase, the epidermal growth factor (EGF) receptor is downregulated in autophosphorylation by wild-type FKBP12 in membrane preparations of A431 cells. The effect is essentially linked to an intact active site, and can be abolished by inhibiting the enzyme with either rapamycin or FK506 (Lopezilasaca et al. 1998). A metastable receptor protein conformation might confer sensitivity to the catalytic effect of FKBP12.

Substrate-like interaction between Cyp18 and the Asn286-Pro287 bond in the ^{15}N-labeled Src homology 2 (SH2) domain of interleukin-2 tyrosine kinase (Itk) have been directly detected by line broadening in the HSQC spectrum in vitro. This tyrosine kinase of the Tec family participates in the intracellular signaling events leading to T cell activation, and plays a role in thymocyte maturation. The SH2 domain controls the catalytic activity of the neighboring nonreceptor protein tyrosine kinase domain intramolecularly, and has some relation to the cause of the hereditary immune disorder, X-linked agammaglobulinemia. Full tyrosine kinase activation requires the interaction between the SH3 and the SH2 domain of Itk. This interaction can be mimicked by self-association of a dual SH3/SH2 domain-containing fragment of Itk. The biologically active, SH3 binding conformer of the isolated SH2 domain is the *cis* Asn286-Pro287 species. Isomer-specific recognition resides in a polyproline-mimicking conformation typical of SH3 ligands that cannot be adopted by the *trans* Asn286-Pro287 conformer (Mallis et al. 2002). However, as in other cases (Bosco et al. 2002; Yuan et al. 1998), low sensitivity of the NMR methods for proteins restricts catalyzed peptide bond *cis/trans* isomerization accessibility to test tube experiments.

An impressive example of PPIase-mediated signaling effects is also seen in the bacterial world.

The lysis gene E of bacteriophage phiX174, which encodes a 92 amino acid membrane protein, provides an example of *cis/trans* isomerization control exerted by the cytoplasmic metal ion-dependent FKBP-like PPIase SlyD of *E. coli*. Western blots and pulse-chase labeling experiments showed that protein E is highly unstable in a *slyD*$^{-}$ background. Perhaps the protein E is able to escape from a protease-sensitive native state isomer by SlyD catalysis. The results indicate that SlyD is required for accumulating the E protein to the levels required to exert its lytic effect (Bernhardt et al. 2002). Elements of enzymatic selectivity could be identified by virtue of an incomplete compensation by other PPIases generated in lysis-inactivating *SlyD* negative strains (Witte et al. 1997).

Peptide bond *cis/trans* isomerases as presenter proteins in immunosuppression

The causal relationship between multiple drug effects and the multifunctional nature of the PPIases existing in blood cells became obvious in transplantation medicine (Schreier et al 1993; Navia et al 1996).

According to the reaction principles discussed in "Reaction profile of peptide bond *cis/trans* isomerases," loss of PPIase function by inhibition of enzyme activity should lead to multisystem failure, if not general toxicity, for example, in lymphocytes. Strong inhibitors for their cognate PPIases are the low-molecular weight microbial immunosuppressants FK506, ascomycin, and rapamycin of the peptidomacrolide type, as well as CsA and sanglifehrin. The unique feature they all share is that they all have numerous effects on cell signaling when administered at doses lower than toxic concentrations. Strikingly, the most prominent effect of CsA and FK506, the prevention of clonal expansion after antigenic stimulation of T cells, has been discussed exclusively in terms of presenter protein strategy (mechanistic principle 3; Vogel et al. 2001). Only recently has attention turned to the role of inhibiting the *cis/trans* isomerase function for the interpretation of drug effects under physiological and pathophysiological conditions (Steiner et al. 1997; Bartz et al. 1995; Christner et al. 2001; Waldmeier et al. 2002; Bennett et al. 1998). Presently, there is no evidence that immunosuppression involves a holding function or proline-directed binding function of PPIases (mechanistic principles 2 and 4). At first glance, the presenter protein mechanism is entirely consistent with structure-function relationships, data on the effective drug/PPIase ratio in lymphocytes, and results of affinity chromatography (Schreiber and Crabtree 1992).

It still remains a challenge to identify the predisposing structural elements behind presenter protein strategies of peptide bond *cis/trans* isomerases. The molecular background, which guides the initiation of the structural changes induced in PPIase/drug complexes, has been delineated. The PPIase forms a binding platform for the drug, with a portion of the drug molecule presented to other components for interaction within the cytosol. Apparently, gain-of-function presentation of peptide chains could cause a molecular interaction profile which is not found with the isolated components.

With this mechanism operative, cytosolic proteins recruited include the protein phosphatase 2B, calcineurin (Cn) for both the Cyp18/CsA and the FKBP12/ FK506 complexes (Liu et al. 1991; Friedman and Weissman 1991), and the mammalian target of rapamycin (mTOR) for the FKBP12/rapamycin complex. mTOR is a member of the phosphoinositide kinase-related kinase family (Brown et al. 1994; Sabers et al. 1995). So far, there is no protein known to be recruited by the Cyp18/sanglifehrin complex.

Keeping in mind the broad efficacy of immunophilin complexes in immunoresponse, there might be other natural molecules the PPIases are likely to be ready to present to other cell constituents. Currently, they are all found in microorganisms, and are only present in mammals under artificial conditions. Thus, the difficulty in interpreting the general significance of the presenter protein strategy for the physiological role of PPIase arises from the fact that there is no example for endogenous molecules acting via presenter PPIases in a gain-of-function mechanism in mammalian cells.

It is now important to reemphasize the strong competitive inhibition of most PPIases depicted in Figs. 5 and 6 by at least one member of the immunosuppressive drugs, while cross-inhibition was not seen with CsA and the peptidomacrolide group of compounds. In addition to their action on other cell types, all inhibitors have in common the ability of downregulating the T lymphocyte activation program at the nanomolar concentration level. Consequently, the drugs can prevent allograft rejection, can be used for treating autoimmune diseases, and are prophylactic for the prevention of graft-versus-host diseases following bone marrow transplantation. However, they do not interfere with the T cell response by blocking common signaling pathways except FK506 (along with the related drug ascomycin) and CsA, which address similarities through targeting the Cn-dependent NFAT transcription pathway. Using that criterion, it is not surprising that side effects encountered under clinical therapy are similar for FK506 and CsA (Mihatsch et al. 1998).

Downregulation of IL-2 expression is thought to play a key role in conferring susceptibility to CsA and FK506 effects on the growth of T-cells involved in cellular immunoresponse.

In contrast, rapamycin inhibits the ability of IL-2 to stimulate the transition of T lymphocytes from G1 to S phase of the cell cycle.

Similar to rapamycin, sanglifehrin blocks IL-2-dependent proliferation in the G1 phase, and decreases the IFN-γ IL-4 but not the IL-2 production of T cells. The proliferation of mitogen-activated B cells is also downregulated, but, unlike rapamycin, sanglifehrin has no effect on CD154/IL-4-induced antibody synthesis (Zenke et al. 2001).

Inactivation of the nuclear factor of activated T cells (NFAT), nuclear factor κB (NFκB) and c-jun terminal kinase during G0 to G1 cell cycle progression, which followed the inhibition of Cn-mediated protein dephosphorylation, is believed to be the primary process targeted by CsA and FK506. However, the drugs on their own do not have the potential to block the enzymatic activity of Cn. When bound to the presenter PPIase of the host cell, the drugs possess inhibitory surfaces complementary to two enzymes, Cn and the respective PPIase. The ternary complexes of FKBP12/FK506/Cn and Cyp18/CsA/Cn, which are dissolved in their three-dimensional structures, provide a structural basis for understanding Cn inhibition by gain-of-function. In these complexes, the interacting surface is represented by atoms from the drug as well as atoms from the presenter protein (Griffith et al. 1995; Jin and Harrison 2002; Huai et al. 2002). FKBP51 demonstrates, unlike FKBP12, affinity to Cn in the absence of FK506. However, overexpression of FKBP51 in Jurkat T cells did not significantly affect the transcriptional activation of NFAT upon stimulation. FK506 effects on NFAT-mediated transcription are not influenced by this interaction. The binding site on Cn for FKBP51 mapped to the tetratricopeptide motif repeats (TPR), a region outside of the FKBP12-like domain (Li et al. 2002). In this case, the binding of GST-FKBP51 to Cn was FK506-independent and independent of a requirement for calcium or exogenous calmodulin. Unfortunately, in many biological studies CsA and FK506 per se were designated as Cn inhibitors, neglecting the proven inactivity of the drugs in in vitro inhibition studies. Since presenter PPIases must be present for Cn inactivation, interpretation of CsA(FK506) effects on physiological signals urgently requires information about the cellular Cyp18(FKBP12) concentration. It came as a surprise to find that CsA derivatives like [Dimethylaminoethylthio-sarcosine]^3-cyclosporin A can indeed inhibit Cn on their own in the low micromolar range, but are unable to inhibit Cn after formation of the [Dimethylaminoethylthio-sarcosine]^3-cyclosporin A /Cyp18 binary complex (R. Baumgraß et al., submitted). Probably the most important reason for the occurrence of gain-of-function mechanisms is alteration in the macrocyclic conformation of the drug when bound to the presenter PPIase. When unbound, CsA differs greatly in conformation from the Cyp18-bound state whereas [Dimethylaminoethylthio-sarcosine]^3-cyclosporin A adopts the gain-of-function conformation already in its unbound state. Composite interaction surfaces might play a minor role for the presenter protein principle. If so, the presenter protein function of FKBPs and cyclophilins is probably restricted to a few macrocyclic ligands and cannot serve as a guide to the physiological function of these enzymes in mammals. In addition, the demonstration of an influence of CsA and FK506 on the signal of any biological assay does not constitute by itself final proof as to the involvement of Cn in cell signaling. Although the use of nonimmunosuppressive CsA(FK506) derivatives has permitted the application of Cn-inactive PPIase inhibitors, corresponding experiments with specific small molecule Cn inhibitors, which are inactive toward PPIases, are missing. In fact, the polyphenolic aldehyde gossypol, the only drug-like inhibitor that exhibits Cn specificity among the protein phosphatases, is inactive toward PPIases. It inhibits NFAT-driven transcription and NFATc1 nuclear translocation in purified human T cells. However, gossypol has already been reported to inactivate a number of other signaling proteins, thus preventing a clear differentiation of Cn-mediated pathways in immunosuppression (Baumgraß et al. 2001).

Unexpectedly, differential cDNA-based microarray analyses, which made use of administration of either CsA or FK506 to detect sensitive early genes in human peripheral blood lymphocytes, only result in the appearance of genes regulated by a common signaling pathway, as the Cn-mediated dephosphorylation represented in the gain-of-function mechanism. There was almost no gene found that is differently expressed in response to the inhibition of either cyclophilins or FKBPs (Cristillo and Bierer 2002). From these facts, there is no evidence that the specific inhibition of PPIases plays a role in immunosuppression except complete redundancy, or when there were isomer-specific posttranslational modifications in immunoresponse. However, the relation of the drug concentration to the high intracellular level of the PPIases is critical for the effects to be expected.

On the other hand, the usual therapeutic levels of CsA in renal transplant patients is associated with only partial inhibition of Cn of about 50–85% reduction of activity in circulating lymphocytes (Batiuk et al. 1995). The Cn fraction resistant to inhibition is even larger for FK506 administration in several tissues (Halloran et al. 1998). Serum CsA level can approximately predict clinical parameters of immunosuppression in individuals, but is poorly correlated with Cn activity of lymphocytes. It was hypothesized that partial Cn inhibition might account for both the immunosuppression and the immunocompetence of CsA-treated patients (Batiuk et al. 1996,1997). More recently, the prevailing assumption that cyclophilins and FKBPs are abundant and not limiting for inhibition of Cn by CsA or FK506 has been challenged. Combined experiments on drug effects on NFATC2 dephosphorylation using western analysis, interferon-γproduction using ELISA, and Cn phosphatase activity using the Cn assay in human peripheral blood leukocytes and mouse spleen cell suspension led to the assumption of the existence of a subset of Cn critical for lymphocyte activation (Kung and Halloran 2002). The nature of this critical Cn form is not yet known. An important consequence of the CsA(FK506)-mediated immunosuppression, which follows from the gain-of-function, is the possibility that Cn depletion in T lymphocytes should cause CsA(FK506) resistance. The gene targeted for deletion, the catalytic subunit CnA, exists in three isoforms (α, β, and γ), with the subunit β as the predominant Cn isoform expressed in lymphocytes (Jiang et al. 1997). After the finding that CnAα (-/-) T cells remained sensitive to both cyclosporin A and FK506 (Zhang et al. 1996) CnAβ (-/-) T became especially important. These cells displayed defective T cell development characterized by fewer total CD3 cells and reduced CD4 and CD8 single positive cells (Bueno et al. 2002). Defective proliferative capacity and decreased IL-2 production in response to PMA/ionomycin and T cell receptor cross-linking were obtained as well. The CnAβ (-/-) mice were also permissive to allogeneic tumor-cell transplantation, similar to cyclosporin-treated wild-type mice. However, enriched CnAβ (-/-) T cells still showed CsA- and FK506-sensitive proliferation in response to CD3 receptor crosslinking and PMA/ionomycin stimulation. Functional redundancy with the CnAα subunit might exist. Despite this concern, a Cn-independent pathway, which is probably related to peptide bond *cis/trans* isomerase inhibition, might become associated with the CsA(FK506)-induced immunosuppression.

Effects of peptide bond *cis/trans* isomerase gene deletion

One of the many examples of the cellular manifestation of the enormous complexity involved in CsA, FK506, and rapamycin administration is found in expression microarrays and proteomics.

Typically, several hundred polypeptides were found to be up- or downregulated when comparing CsA-treated and untreated ConA-activated T-lymphocytes (Truffa-Bachi et al. 2000). The modulation of transcription by rapamycin in *S. cerevisiae* causes the activation or repression of 297 genes more than fourfold (Hardwick et al. 1999). As discussed above, phenotypic information from the application of these inhibitors in in vivo experiments cannot be utilized to draw reliable conclusions about the normal and the pathophysiological function of a particular peptide bond *cis/trans* isomerase in whole organisms.

To circumvent the current problems, the potential of phenotypic information from gene disruption has been used. Other than inhibitor effects on the phenotype of an already existing living system, gene deletion experiments include massive growth selection. Thus, results of inhibitor studies and gene disruption experiments must not necessarily coincide.

Some of the findings and conclusions derived from the analyses of organisms deficient of individual peptide-bond *cis/trans* isomerase genes are summarized in Table 2.

Most strikingly, all genes encoding cyclophilins and FKBPs, a total of 12, can be deleted without seriously affecting viability in *S. cerevisiae* (Dolinski et al. 1997). The only PPIase that remained is the pSer/pThr-Pro-specific Ess1, a parvulin-like enzyme that proved to be essential in yeast. Consequently, disruption of the single PPIase genes, except Ess1, does not have remarkable phenotypic changes under optimal growth conditions. Nevertheless, effects have been found for single gene deletion if special functions were requested. CPR1 (Cyp17) gene dependence was conferred by mutations in ZPR1, a gene encoding an essential zinc finger protein. CPR1 dependence was suppressed by overexpression of the Zpr1p binding translation factor EF1α, Cyp40, or FKBP12 (Ansari et al. 2002). Functional redundancy has already been discussed in numerous in vivo studies on PPIases, but is now obvious in these experiments, and in others (Fujimori et al. 2001; Arevalo-Rodriguez et al. 2000). Kinetic studies of the protein transport indicated that Zpr1p accumulates in the nucleus in the absence of PPIase-active Cyp17 because the nuclear export of

Table 2 Gene deletions

Enzyme	Organism	Gene	Viability[a]	Physiological defects	Reference
Cyclophilins					
Cyp18	Mice Jurkat T cells	*Ppia*	+	Prototypic FKBP, partially HIV-1 resistant human T cells, no growth effects on T cells	Colgan et al. 2000; Braaten and Luban 2001
Cyp	*Aspergillus nidulans*	*cypB*	+	Not essential for vegetative growth	Joseph et al. 1999
Cyp17	*Saccharomyces cerevisiae*	CPR1	+	Primary receptor for CsA, becomes essential if ESS1 is compromised, decreased heat shock survival, compromised nuclear transport of Zpr1p zinc finger protein	Arevalo-Rodriguez et al. 2000; Sykes et al. 1993; Ansari et al. 2002
Cyp	*Saccharomyces cerevisiae*	Many	+	A dodecuplet mutant strain (CPR1-CPR8, FPR1-FPR4) is viable, small growth defect attributable to CPR7 (Cyp40 homolog) disruption	Dolinski at al 1997
Cyp18	*Cryptococcus neoformans*	*CPA1, CPA2*	+	cpa1,cpa2 double mutant inviable and avirulent at 30°C and 39°C and are unilaterally sterile. Single gene mutations less severe	Wang et al. 2001
Cyp19	*Erwinia chrysanthemi*	*rotA*	+		Pissavin et al. 1997
Cyp15	*Cyanobacterium synechococcus*	*rotA*	–	Essential for growth	Hassidim et al. 1992
Cyp20	*Acinetobacter calcoaceticus*	*rotA*	+		Kok et al. 1994
Cyp21	*Escherichia coli*	*rotA*	+		Kleerebezem et al. 1995
Cyp17	*Bacillus subtilis*	*ppiB*	+	*tig, ppiB* double deletion reduced starvation viability	Herrler et al. 1994; Gothel et al. 1998
FKBP					
FKBP36	Human	*FKBP6*	+	Deleted in Williams syndrome	Meng et al. 1998
FKBP12	Mice	*fkbp12*	+	Prototypic FKBP, severe dilated cardiomyopathy, septal defects, neural tube closure, cell cycle arrest in G1 phase	Shou et al.1998; Bassing et al. 1998; Aghdasi et al. 2001
FKBP12.6	Mice	*fkbp12.6*	+	Estrogen-dependent cardiac hypertrophy in male mice, dysregulation of calcium release	Xin et al. 2002
FKBP12	*Saccharomyces cerevisiae*	*FPR1*	+	Slow growth, primary FK506 receptor	Heitman et al. 1991; Wiederrecht et al. 1991

Table 2 (continued)

Enzyme	Organism	Gene	Viability[a]	Physiological defects	Reference
FKBP12	*Schizosaccharomyces pombe*	*fkh1*	+	No effect on vegetative growth, mating deficiency	Weisman et al. 2001
FKBP42	*Arabidopsis thaliana*	*twd1*	+	Dwarfism, helical root and stem rotation	Perez-Perez et al. 2001; Kamphausen et al. 2002
FKBP70	*Arabidopsis thaliana*	*pas1*	+	Defects in cell proliferation and cell division	Vittorioso et al. 1998
FKBP13	*Neurospora crassa*	*fkbp*	+	resistant toward FK506, *fkbp,cyp* double mutant viable	Barthelmess and Tropschug 1993
FKBP	*Salmonella typhimurium*	*fkpA*	+	Attenuation of virulence, reduced intracellular multiplication in macrophages and Caco-2 cells	Horne et. al. 1997
FKBP25	*Legionella pneumophila*	*mip*	+	Attenuation of virulence, reduced intracellular multiplication	Cianciotto and Fields 1992
FKBP29	*Aeromonas hydrophila*	*fkpA*	+	No effect on virulence	Wong et al. 1997
FKBP21	*E. coli*	*slyD*	+	Essential host protein for bacteriophages to lyse, loss of phage E protein stability	Roof et al. 1994
FKBP29	*E. coli*	*fkpA*	+	Synthetic lethality in *skp fkpA* double mutants	Horne and Young 1995; Dartigalongue et al. 2001
Trigger factors					
Trigger factor	*E. coli*	*tig*	+	Enhanced filamentation, decreased cold viability *tig dnak* double mutation is lethal	Guthrie and Wickner 1990; Teter et al. 1999; Deuerling et al. 1999; Kandror and Goldberg 1997
Trigger factor	*Bacillus subtilis*	*tig*	+	Growth retardation in minimal media, *tig, ppiB* double deletion reduced starvation viability	Gothel et al. 1998
Trigger factor	*Actinobacillus pleuropneumoniae*	*tig*	+	Attenuated virulence, reduced incidence of fatal fibrinous hemorrhagic bronchopneumonia in swine	Fuller et al. 2000
Trigger factor	*Streptococcus pyogenes*	*ropA*	+	Maturation defects on SpeB protease	Lyon et al. 1998
Parvulins					
Pin1	Mice	*pin1*	+	Fibroblasts deficient in restart of proliferation, decreased body size, testicular atrophy	Fujimori et al. 1999; Liou et al. 2002

Table 2 (continued)

Enzyme	Organism	Gene	Viability[a]	Physiological defects	Reference
Dodo	*Drosophila melanogaster*	*dodo*	+	Oogenic defects	Maleszka et al. 1996; Hsu et al. 2001
PTF1/ESS1	*Saccharomyces cerevisiae*	*ESS1*	–	Essential for growth, involved in RNA 3′-end processing	Hani et al. 1995
Pin1(+)	*Schizosaccharomyces pombe*	*pin1(+)*	+	Increased UV sensitivity, several synthetic defects	Huang et al. 2001
Par10	*E. coli*	*parA*	+	Prototypic parvulin, reduced viability to oxidative stress	B. Hernandez-Alvarez, unpublished data
ESS1	*Candida albicans*	*ESS1*	–	Essential for growth, morphogenetic defects	Devasahayam et al. 2002
SurA	*E. coli*	*surA*	+	Defective cell envelope, essential for growth in the stationary phase	Tormo et al. 1990
PpiD	*E. coli*	*ppiD*	+	Defective membrane protein profile, *PpiD,SurA* double mutant causes synthetic lethality	Dartigalongue and Raina 1998
SurA	*Salmonella enterica*	*surA*	+	Defective in the ability to adhere and invade eukaryotic cells	Sydenham et al. 2000
PrsA	*Bacillus subtilis*	*prsA*	–	Essential function in protein secretion	Kontinen and Sarvas 1993
hsp70					
DnaK	*E. coli*	*dnaK*	+	Confers heat shock sensitivity, prevents protein degradation	Paek and Walker 1987; Hesterkamp and Bukau 1998; Sherman and Goldberg 1992
DnaK	*Bacillus subtilis*	*dnaK*	+	Confers heat shock sensitivity, *tig*, *dnaK* double mutant cause synthetic lethality at 53°C	Reyes and Yoshikawa 2002

[a] Viability under normal growth conditions

Zpr1p was defective in the mutant cells. Considering the different substrate specificities (Yaffe at al. 1997), complete functional overlap between Ess1 and FKBPs/Cyps appears to be improbable. Partial overlap has been detected by a Cyp17-mediated suppression of the deadly effect of *ess1* conditional and null mutations. The level of enzyme activity of Cyp17 active site variants correlates with the suppression of the *ess1* null mutation (Arevalo-Rodriguez et al. 2000).

It is highly improbable given the remarkable phylogenetic conservation and the high number of different members that the Cyps and FKBPs do not affect important metabolic pathways in higher eukaryotes. Furthermore, archetypical cyclophilin genes have some characteristics typical of housekeeping genes that encode proteins required for general cellular functions (Colgan et al. 2000). On the other hand, the expression of certain peptide bond *cis/trans* isomerases has been shown to be upregulated under various stress conditions (Marivet et al. 1992; Andreeva et al. 1997; Robson et al. 1999; Kurek et al. 1999). According to this stress response, which might affect functional compensation for deleted genes, a specific physiological state of the cell utilized to study gene deletion has to be considered. In other words, living cells may respond differently to reduced PPIase activity depending upon the experimental conditions. Therefore, it is noteworthy that replication of a primary isolate HIV-1 exhibits a larger functional dependence on Cyp18 than the replication of a laboratory-adapted strain does (Braaten and Luban 2001).

By contrasting the results with *Saccharomyces cerevisiae*, cyclophilin gene disruption in the pathogenic fungus *Cryptococcus neoformans* gives impressive phenotypic expressions (Wang et al. 2001). Two prototypic cyclophilin members that share 93% sequence identity in the 163 amino acids encoded by the *CPA1* and *CPA2* genes were removed, either separately or in tandem. The sequence variation of *CPA1* and *CPA2* results from amino acid substitutions in the N-terminal part of the proteins that does not seem to define active site residues. The single mutation in *cpa1* already caused defects in growth at 39°C, indicating partially overlapping functions of both cyclophilins.

The cpa1 cpa2 double mutants exhibited severe synthetic defects in growth and virulence. $CPA1^{Arg55Ala}$ active site mutant restored growth to cpa1 cpa2 double mutants at ambient but not at higher temperatures, suggesting that the prolyl isomerase activity of the cyclophilin is crucial to their in vivo function. Summarizing the defects of the mutant strains, a strongly reduced ability to respond to stress conditions was noted.

Presently, no simple explanation can be offered for the entirely different findings with PPIase gene disruption strains of *Saccharomyces cerevisiae* and *Cryptococcus neoformans.*

Additional studies reflect a general relationship between the expression profile of peptide bond *cis/trans* isomerases and virulence. Although Cyp18 does not play an influential role for the growth rate of Jurkat T cells, the onset of HIV-1 virus replication is delayed considerably in the $PPIA^{-/-}$ cells. Cyp18 proved to be the only host cell cyclophilin associated with HIV-1 virulence. The effect is specific to HIV-1 because HIV-2 replication is insensitive to the genetically modulated level of Cyp18, and to the presence of CsA-like Cyp18 inhibitors. Since the $PPIA^{-/-}$ T cells are compatible with life, an enhanced resistance to HIV-1 transmission for individuals carrying a low expression level of Cyp18 was hypothesized (Braaten and Luban 2001).

The Mip (macrophage infectivity potentiator) protein represents secreted dimeric FKBP25 of *Legionella pneumophila*, which is necessary for optimal surmounting intracellular stress conditions applied to the bacteria in macrophage-like cell line U937, human blood monocytes, and *Acanthamoeba castellanii.* By adopting a special type of global fold, which is shared by another folding catalyst, the protein disulfide bond isomerase DsbC, dimeric FKBPs from bacteria can achieve concerted action in two catalytic domains on a substrate (McCarthy et al. 2000; Pereira et al. 2002; Riboldi-Tunnicliffe et al. 2001). Genes encoding the wild-type FKBP25 and a panel of site-directed mutant enzymes were used to complement three different Mip-negative mutants of *Legionella pneumophila* (Corby, Philadelphia I, and Wadsworth) for survival in different host cells. Infection in cell culture with FKBP25-negative mutant bacteria was reduced approximately 50–100-fold in *Acanthamoeba castellanii* and in human mononuclear phagocytes compared with their isogenic FKBP25-positive parental strains. Unexpectedly, the wild-type invasion rate could be restored by introducing an intact copy of *mip* genes into FKBP25-negative strains that encode for enzymes with a greatly reduced PPIase activity (Wintermeyer et al. 1995). The functional significance of the complementation experiments is not yet clear because of the existence of residual enzyme activity in the FKBP25 variants. Generally, completely inactive peptide bond *cis/trans* isomerase variants retaining the native-like fold have not yet been observed.

A further PPIase for which in vivo function has been examined genetically in *E.coli* is the parvulin-like enzyme SurA. Cellular properties thought to be related to important *E.coli* survival functions, $E\sigma^E$ activity, and sensitivity to membrane perturbants have been investigated by gene complementation experiments with SurA fragments and active site variants of SurA in *surA*⁻ cells. Although complementation experiments with a variety of domain constructs gave different results in the two assays, the dispensability of both parvulin domains as well as of the PPIas2/26/03e activity for facilitating porin maturation was suggested (Behrens et al. 2001).

Another periplasmic PPIase carrying a single parvulin-like domain, the PpiD, was shown to be responsible for membrane defects. Phenotypically, there is some overlap between *ppiD* and *surA* null mutants. The combined deletion of both parvulin-like enzymes is synthetically lethal (Dartigalongue and Raina 1998). The rescue of membrane leakiness upon overexpression of PpiD protein in the *surA*⁻ background indicates functional overlap as well. Interestingly, ppiD genes carrying point mutations in a region encoding for the active site of the PPIase cannot complement the *surA*⁻ cells for loss of function. Consequently, the corresponding protein variants were shown to exhibit a greatly reduced level of PPIase activity. The divergent results obtained with the complementation of surA⁻ strains are striking.

A major difference between both enzymes exists regarding the level of PPIase activity in the standard tetrapeptide assay. Enzyme activity is very low for SurA when compared to PpiD. Estimation of the residual PPIase activity of enzyme variants generated by site-directed mutagenesis remains a difficult problem when activity does not markedly exceed the background level. It would be interesting to see whether there is a possibility of crossing a *ppiD* null allele into a strain that carries the PPIase inactive SurA variants on a plasmid.

In higher organisms, the deletion of single PPIases genes often causes severe growth defects.

The only FKBP isoforms whose in vivo function in mammals has been examined genetically are FKBP12 and FKBP12.6. *FKBP12*$^{-/-}$ mice suffer early embryonic death caused by cardiac defects. The severity of the phenotype and the cause of death was examined in detail (Shou et al. 1998). A relationship between FKBP12 and transforming growth factor-

β (TGF-β)-mediated transcriptional activation was suggested from the involvement of this PPIases in the TGF-β receptor-I assembly.

Using *FKBP12*$^{-/-}$ and *FKBP12*$^{+/+}$ fibroblasts in conjunction with FKBP12 inhibitor effects, a unique physiological role of FKBP12 in TGF-β signaling in primary fibroblasts and thymocytes could not be observed (Bassing et al. 1998). More recently, however, fibroblasts from *FKBP12* $^{-/-}$ mice were found to suffer from cell cycle arrest in G1 phase which could be rescued by wtFKBP12 transfection. Accordingly, mutant mice grew to only a small size. This arrest is mediated by marked augmentation of p21(WAF1/CIP1) levels, which cannot be further augmented by TGF-β. The p21 upregulation and cell cycle arrest derives from hyperactive TGF-β receptor signaling, which is normally inhibited by FKBP12 (Aghdasi et al. 2001). The heterodimeric TGF-β receptor consists of the serine/threonine kinase subunits type I and type II (TβR-I and TβR-II). FKBP12 binds to the Leu193-Pro194 segment of the juxtamembrane portion of the unactivated TβR-I, capping the TβRII phosphorylation site (Huse et al. 1999). Loss of the FKBP12 binding properties in TβRI variants lacking this proline-directed binding site accompanies receptor hypersensitivity resulting in leaky signaling (Chen et al. 1997). Consequently, cell cycle arrest in the *FKBP12*$^{-/-}$ mouse fibroblasts is prevented by transfection with a dominant-negative TGF-β receptor construct.

FKBP12 and FKBP12.6 bind to skeletal and cardiac ryanodine receptors, respectively, and regulate the release of calcium from intracellular stores. Heart cells from *FKBP12.6* $^{-/-}$ mice in particular show an enhanced calcium-induced calcium release from the sarcoplasmic reticulum, indicating a role of FKBP12.6 in ryanodine receptor (RyR2) gating (Xin et al. 2002). Similarly, when exposed to of 0.01–10 μM FK506, coronary arterial smooth muscle cell preparations show a significantly increased open probability of reconstituted RyR2/calcium release channels of the sarcoplasmic reticulum, which relates to the inhibition of FKBP12.6. Apparently, FKBP12.6 is also involved in the enhancing effects of the second messenger cADP ribose on RyR2 channel opening. However, only male mice developed an enlarged heart because estrogen is a protective factor in suppressing the response to calcium dysregulation in the heart cells of female mice (Tang et al. 2002).

From the seventeen FKBP-like proteins encoded in the genome of *Arabidopsis thaliana*, the nuclear-localized FKBP70 and the FKBP42 homologs were targeted by gene disruption (Vittorioso et al. 1998; Perez-Perez et al 2001). These *pas1* (FKBP70) mutant plants cannot grow normally because phenotypes show ectopic cell proliferation in cotyledons, extra layers of cells in the hypocotyl, and an abnormal apical meristem. Flowers were abnormal and sterile. It was suggested that this FKBP controls plant cell division and differentiation in a way similar to antioncogenes in mammals (Harrar et al. 2001). Disruption of the *twd1* gene that encodes for FKBP42, leads to multiple morphogenic defects (Perez-Perez et al. 2001). The domain structure of FKBP42 suggests the involvement of a heteroligomeric receptor complex that requires hsp90 to function (Kamphausen et al. 2002).

The deletion of the APIase DnaK (and its cochaperone DnaJ) in the Δ*tig* background is lethal in *E. coli* (Teter et al. 1999; Deuerling et al. 1999) because a large fraction of the newly synthesized polypeptides cannot be properly folded in time. Functionally, DnaK and trigger factor share some common features. Both demonstrate considerable affinity to unfolded polypeptide chains and exhibit peptide bond *cis/trans* isomerase activity. Trigger factor differs from DnaK in that its enzyme activity might be restricted to the catalysis of prolyl bonds, it does not display affinity to cochaperones, and it does not contain an AT-

Pase function. Using trigger factor variants, diminished cold-stress tolerance of Δ*tig* *E. coli* could not be rescued with activity-depleted variants in transfection experiments (Schiene-Fischer et al. 2002b). Interestingly, the freezing tolerance of DnaK overexpressing *E. coli* strains and the diminished recovery from frozen storage of the F'l⁻ Δ*dnaK*52::Cmr strain are reminiscent of trigger factor effects (Chow and Tung 1998). DnaK variants with attenuated APIase activity are still lacking, thus preventing the determination of the role of this enzyme activity by complementation experiments.

It is noteworthy that studies often indicate minor phenotypic consequences in response to multiple disruptions of PPIase genes in microorganisms. However, higher organisms are seriously affected even in response to a single gene deletion. Currently, knowledge of the involvement of the peptide bond *cis/trans* isomerization in the dramatic phenotypic changes of PPIase-deficient higher organisms is very limited. Dissecting ground state affinity effects from catalytic processes urgently requires complementation experiments with PPIase-deficient enzyme variants (Fischer et al. 1998).

Conclusions

Peptide bond *cis/trans* isomerases have evolved to allow enzyme control of switching backbone conformations in native proteins, folding intermediates, and unfolded chains. In addition, the intrinsic property of peptide bond *cis/trans* isomerases to decrease the internal barrier to rotation enhances the rotational flexibility around the targeted C-N bond, which in turn leads to backbone "liquification" localized to a "hot spot" protein region. Two major families, the PPIases and the APIases, form the enzyme class. However, expansion of the family numbers continues to occur. A single peptide bond *cis/trans* switch was often found to separate biologically active from inactive polypeptides. Peptide bond *cis/trans* isomerases play a key role in the spatial-temporal control of these forms. Other biochemical properties play an auxiliary role, and might perform more specialized physiological functions. Indirect methods have been used to demonstrate the involvement of catalyzed peptide bond *cis/trans* isomerization of proteins in cell signaling because direct methods only rarely exist. In contrast, the auxiliary properties of a peptide bond *cis/trans* isomerase proved to be more easily detected in vivo and in vitro.

Experiments combining gene deletion with variant enzyme gene transfection, and the pharmacological approach with monofunctional enzyme inhibitors, were applied most successfully. Nevertheless, specific inhibitors that can differentiate among the isoenzymes of the subfamilies are still missing. Experimental tools for the direct monitoring of the isomeric state of a particular peptide bond in a protein in situ are still lacking. Achievement of this ultimate goal, however, is important but must await further research.

Future directions in the research on accelerated conformational switching will be based on the emerging picture of catalyzed folding events in proteins that govern the spatial and temporal control of bioactiviy in cell signaling.

Acknowledgments. We are indebted to C. Schiene-Fischer for providing figures and for many valuable comments and constructive discussions on the manuscript. Many thanks to F. Erdmann and A. Hemming for supplying Fig. 4 and for critical reading of the manuscript. We gratefully acknowledge the financial support of the Fonds der Chemischen Industrie to G.F.

References

Aghdasi B, Ye K, Q, Resnick A, Huang A, Ha HC, Guo X, Dawson TM, Dawson VL, Snyder SH (2001) FKBP12, the 12-kDa FK506-binding protein, is a physiologic regulator of the cell cycle. Proc Natl Acad Sci USA 98:2425–2430

Allain F, Vanpouille C, Carpentier M, Slomianny MC, Durieux S, Spik G (2002) Interaction with glycosaminoglycans is required for cyclophilin B to trigger integrin-mediated adhesion of peripheral blood T lymphocytes to extracellular matrix. Proc Natl Acad Sci USA 99:2714–2719

Andreeva L, Motterlini R, Green CJ (1997) Cyclophilins are induced by hypoxia and heat stress in myogenic cells. Biochem Biophys Res Commun 237:6-9

Ansari H, Greco G, Luban J (2002) Cyclophilin a peptidyl prolyl isomerase activity promotes ZPR1 nuclear export. Mol Cell Biol 22:6993–7003

Arevalo-Rodriguez M, Cardenas ME, Wu XY, Hanes SD, Heitman J (2000) Cyclophilin A and Ess1 interact with and regulate silencing by the Sin3-Rpd3 histone deacetylase. EMBO J 19:3739–3749

Arie JP, Sassoon N, Betton JM (2001) Chaperone function of FkpA, a heat shock prolyl isomerase, in the periplasm of *Escherichia coli*. Mol Microbiol 39:199–210

Asea A, Kraeft SK, Kurt-Jones EA, Stevenson MA, Chen LB, Finberg RW, Koo GC, Calderwood, SK (2000) HSP70 stimulates cytokine production through a CD14-dependant pathway, demonstrating its dual role as a chaperone and cytokine. Nat Med 6:435–442

Bächinger HP (1987) The influence of peptidyl prolyl *cis/trans* isomerase on the in vitro folding of type III collagen. J Biol Chem 262:17144–17148

Baine P (1986) Comparison of rate constants determined by two-dimensional NMR spectroscopy with rate constants determined by other NMR techniques. Magn Reson Chem 24:304–307

Balbach J, Steegborn C, Schindler T, Schmid FX (1999) A protein folding intermediate of ribonuclease T1 characterized at high resolution by 1D and 2D real-time NMR spectroscopy. J Mol Biol 285:829–42

Barthelmess IB, Tropschug M (1993) FK506-binding protein of *Neurospora crassa* (NcFKBP) mediates sensitivity to the immunosuppressant FK506 resistant mutants identify two loci. Curr Genet 23:54–58

Bartz SR, Hohenwalter E, Hu MK, Rich DH, Malkovsky M (1995) Inhibition of human immunodeficiency virus replication by nonimmunosuppressive analogs of cyclosporine A. Proc Natl Acad Sci USA 92:5381–5385

Bassing CH, Shou WN, Muir S, Heitman J, Matzuk MM, Wang XF (1998) FKBP12 is not required for the modulation of transforming growth factor β receptor I signaling activity in embryonic fibroblasts and thymocytes. Cell Growth Differ 9:223–228

Batiuk TD, Kung LN, Halloran PF (1997) Evidence that calcineurin is rate-limiting for primary human lymphocyte activation. J Clin Invest 100:1894–1901

Batiuk TD, Pazderka F, Halloran PF (1995) Calcineurin activity is only partially inhibited in leukocytes of cyclosporine-treated patients. Transplantation 59: 1400–1404

Batiuk TD, Urmson, J, Vincent D, Yatscoff RW, Halloran PF (1996) Quantitating immunosuppression. Estimating the 50% inhibitory concentration for in vivo cyclosporine in mice. Transplantation 61:1618–1624

Baumgraß R, Weiwad M, Erdmann F, Liu JO, Wunderlich D, Grabley S, Fischer G (2001) Reversible inhibition of calcineurin by the polyphenolic aldehyde gossypol. J Biol Chem 276:47914–47921

Behrens S, Maier R, de Cock H, Schmid FX, Gross CA (2001) The SurA periplasmic PPIase lacking its parvulin domains functions in vivo and has chaperone activity. EMBO J 20:285–294

Bennett PC, Singaretnam LG, Zhao WQ, Lawen A, Ng KT (1998) Peptidyl prolyl *cis/trans* isomerase activity may be necessary for memory formation. FEBS Lett 431:386–390

Bernhardt TG, Roof WD, Young R (2002) The *Escherichia coli* FKBP-type PPIase SlyD is required for the stabilization of the E lysis protein of bacteriophage phi X174. Mol Microbiol 45:99–108

Binsch G (1969) A unified theory of exchange effects on nuclear magnetic resonance line shapes. J Am Chem Soc 91: 1304–1309

Bitto E, McKay DB (2002) Crystallographic structure of SurA, a molecular chaperone that facilitates folding of outer membrane proteins. Structure 10:1489–1498

Bosco DA, Eisenmesser EZ, Pochapsky S, Sundquist WI, Kern D (2002) Catalysis of *cis/trans* isomerization in native HIV-1 capsid by human cyclophilin A. Proc Natl Acad Sci USA 99:5247–5252

Bose S, Weikl T, Bugl H, Buchner J (1996) Chaperone function of hsp90-associated proteins. Science 274:1715–1717

Braaten D, Luban J (2001) Cyclophilin A regulates HIV-1 infectivity, as demonstrated by gene targeting in human T cells. EMBO J 20:1300–1309

Brandts JF, Halvorson HR, Brennan M (1975) Consideration of the possibility that the slow step in protein denaturation reactions is due to *cis/trans* isomerism of proline residues. Biochemistry 14:4953–4963

Brown EJ, Albers MW, Shin TB, Ichikawa K, Keith CT, Lane WS, Schreiber SL (1994) A mammalian protein targeted by G1-arresting rapamycin-eceptor complex. Nature 369:756–758

Bueno OF, Brandt EB, Rothenberg ME, Molkentin JD (2002) Defective T cell development and function in calcineurin A β-deficient mice. Proc Natl Acad Sci USA 99:9398–9403

Bukrinsky MI (2002) Cyclophilins: unexpected messengers in intercellular communications. Trends Immunol 23:323–325

Burbaum JJ, Raines RT, Albery WJ, Knowles JR (1989) Evolutionary optimization of the catalytic effectiveness of an enzyme. Biochemistry 28:9293–9305

Capano M, Virji S, Crompton M (2002) Cyclophilin-A is involved in excitotoxin-induced caspase activation in rat neuronal B50 cells. Biochem J 363:29–36

Carpentier M, Allain F, Slomianny MC, Durieux S, Vanpouille C, Haendler B, Spik G (2002) Receptor type I and type II binding regions and the peptidyl prolyl isomerase site of cyclophilin B are required for enhancement of T-lymphocyte adhesion to fibronectin. Biochemistry 41:5222–5229

Carter P, Wells JA (1987) Engineering enzyme specificity by "substrate-assisted catalysis" Science 237:394–399

Chen YG, Liu F, Massague J (1997) Mechanism of TGF-β receptor inhibition by FKBP12. EMBO J 16:3866–3876

Chiti F, Taddei N, Giannoni E, van Nuland NAJ, Ramponi G, Dobson CM (1999) Development of enzymatic activity during protein folding—detection of a spectroscopically silent native-like intermediate of muscle acylphosphatase. J Biol Chem 274:20151–20158

Chow KC, Tung WL (1998) Overexpression of dnaK/dnaJ and groEL confers freeze tolerance to *Escherichia coli*. Biochem Biophys Res Commun 253:502–505

Christner C, Herdegen D, Fischer G (2001) FKBP ligands as novel therapeutics for neurological disorders. Mini Rev Med Chem 1:377–397

Cianciotto NP, Fields BS (1992) *Legionella-Pneumophila*-MIP gene potentiates intracellular infection of protozoa and human macrophages. Proc Natl Acad Sci USA 89:5188–5191

Clarke SJ, McStay GP, Halestrap AP (2002) Sanglifehrin A acts as a potent inhibitor of the mitochondrial permeability transition and reperfusion injury of the heart by binding to cyclophilin-D at a different site from cyclosporin A. J Biol Chem 277:34793–34799

Colgan J, Asmal M, Luban J (2000) Isolation, characterization and targeted disruption of mouse Ppia: Cyclophilin A is not essential for mammalian cell viability. Genomics 68:167–178

Counterman AE, Clemmer DE (2002) *Cis/trans* signatures of proline-containing tryptic peptides in the gas phase. Anal Chem 74:1946–1951

Cristillo AD, Bierer BE (2002) Identification of novel targets of immunosuppressive agents by cDNA-based microarray analysis. J Biol Chem 277:4465–4476

Dartigalongue C, Missiakas D, Raina S (2001) Characterization of the *Escherichia coli* sigma E regulon. J Biol Chem 276:20866–20875

Dartigalongue C, Raina S (1998) A new heat shock gene *PpiD*, encodes a peptidyl prolyl isomerase required for folding of outer membrane proteins in *E. coli*. EMBO J 17:3968–3980

Davis JM, Boswell BA, Bachinger HP (1989) Thermal stability and folding of type IV procollagen and effect of peptidyl-prolyl cis-trans-isomerase on the folding of the triple helix. J Biol Chem 264:8956–8962

de Crouy-Chanel A, Kohiyama M, Richarme G (1999) Interaction of DnaK with native proteins and membrane proteins correlates with their accessible hydrophobicity. Gene 230:163–170

Deuerling E, Schulze-Specking A, Tomoyasu T, Mogk A, Bukau B (1999) Trigger factor and DnaK cooperate in folding of newly synthesized proteins. Nature 400:693–696

Devasahayam G, Chaturvedi V, Hanes SD (2002) The ess1 prolyl isomerase is required for growth and morphogenetic switching in *Candida albicans*. Genetics 160:37–48

Dobson CM, Sali A, Karplus M (1998) Protein folding—a perspective from theory and experiment. Angew Chem Intern Ed Engl 37:868–893

Dolinski K, Muir S, Cardenas M, Heitman J (1997) All cyclophilins and FK506 binding proteins are, individually and collectively, dispensable for viability in *Saccharomyces cerevisiae*. Proc Natl Acad Sci USA 94:13093–13098

Enghild JJ, Valnickova Z, Thogersen IB, Pizzo SV (1994) Complexes between serpins and inactive proteinases are not thermodynamically stable but are recognized by serpin receptors. J Biol Chem 269:20159–20166

Evans PA, Kautz RA, Fox RO, Dobson CM (1989) A magnetization-transfer nuclear magnetic resonance study of the folding of staphylococcal nuclease. Biochemistry 28:362–70

Fischer G (1994a) Peptidyl prolyl *cis/trans* isomerases and their effectors. Angew Chem Intern Ed Engl 33:1415–1436
Fischer G, Bang H, Berger E, Schellenberger A (1984a) Conformational specificity of chymotrypsin toward proline-containing substrates. Biochim Biophys Acta 791:87–97
Fischer G, Bang H, Mech C (1984b) Determination of enzymatic catalysis for the *cis/trans* isomerization of peptide bonds in proline-containing peptides. Biomed Biochim Acta 43:1101–1111
Fischer G, Heins J, Barth A (1983) The conformation around the peptide bond between the P1- and P2-positions is important for catalytic activity of some proline-specific proteases. Biochim Biophys Acta 742:452–462
Fischer G, Schutkowski M, Küllertz G (2001) Peptidic substrates for directly assaying enzyme activities. Pat. No. DE 100 23743 A1
Fischer G, Tradler T, Zarnt T (1998) The mode of action of peptidyl prolyl *cis/trans* isomerases in vivo: binding versus catalysis. FEBS Lett 426:17–20
Fischer G, Wittmann Liebold B, Lang K, Kiefhaber T, Schmid FX (1989) Cyclophilin and peptidyl-prolyl *cis/trans* isomerases are probably identical proteins. Nature 337:476–478
Fischer G, Wöllner S, Schönbrunner R, Scherer G (1994b) Inhibitors for Peptidyl prolyl *cis/trans* isomerases. Proceedings of the 5th Akabori Conference, Max-Planck Society, Dresden, pp 142–145
Franke EK, Luban J (1996) Inhibition of HIV-1 replication by cyclosporin A or related compounds correlates with the ability to disrupt the gag-cyclophilin A interaction. Virology 222:279–282
Freeman BC, Toft DO, Morimoto RI (1996) Molecular chaperone machines—chaperone activities of the cyclophilin Cyp-40 and steroid receptor associated protein p24. Science 274:1718–1720
Friedman J, Weissman I (1991) Two cytoplasmic candidates for immunophilin action are revealed by affinity for a new cyclophilin—one in the presence and one in the absence of CsA. Cell 66:799–806
Frydman J (2001) Folding of newly translated proteins in vivo: the role of molecular chaperones. Ann Rev Biochem 70:603–647
Fujimori F, Gunji W, Kikuchi J, Mogi T, Ohashi Y, Makino T, Oyama A, Okuhara K, Uchida T, Murakami Y (2001) Crosstalk of prolyl isomerases, Pin1/Ess1, and cyclophilin A. Biochem Biophys Res Commun 289:181–190
Fujimori F, Takahashi K, Uchida C, Uchida T (1999) Mice lacking Pin1 develop normally, but are defective in entering cell cycle from G(0) arrest. Biochem Biophys Res Commun 265:658–663
Fuller TE, Martin S, Teel JF, Alaniz GR, Kennedy MJ, Lowery DE (2000) Identification of *Actinobacillus pleuropneumoniae* virulence genes using signature-tagged mutagenesis in a swine infection model. Microb Pathog 29:39–51
Furutani M, Ideno A, Iida T, Maruyama T (2000) FK506 binding protein from a thermophilic archaeon, *Methanococcus thermolithotrophicus*, has chaperone-like activity in vitro. Biochemistry 39:453–462
Gamble TR, Vajdos FF, Yoo SH, Worthylake DK, Houseweart M, Sundquist WI, Hill CP (1996) Crystal structure of human cyclophilin A bound to the amino-terminal domain of HIV-1 capsid. Cell 87:1285–1294
Garcia-Echeverria C, Kofron JL, Kuzmic P, Kishore V, Rich DH (1992) Continuous flourometric direct (uncoupled) assay for peptidyl prolyl *cis/trans* isomerases. J Am Chem Soc 114:2758–2759
Garcia-Echeverria C, Kofron JL, Kuzmic P, Rich DH (1993) A continuous spectrophotometric direct assay for peptidyl prolyl *cis/trans* isomerases. Biochem Biophys Res Commun 191:70–75
Gothel SF, Scholz C, Schmid FX, Marahiel MA (1998) Cyclophilin and trigger factor from *Bacillus subtilis* catalyze in vitro protein folding and are necessary for viability under starvation conditions. Biochemistry 37:13392–13399
Griffith JP, Kim JL, Kim EE, Sintchak MD, Thomson JA, Fitzgibbon MJ, Fleming MA, Caron PR, Hsiao K, Navia MA (1995) X-ray structure of calcineurin inhibited by the immunophilin-immunosuppressant FKBP12-FK506 complex. Cell 82:507–522
Guthrie B, Wickner W (1990) Trigger factor depletion or overproduction causes defective cell division but does not block protein export. J Bacteriol 172:5555–5562
Halestrap AP, McStay GP, Clarke SJ (2002) The permeability transition pore complex: another view. Biochimie 84:153–166
Halloran PF, Kung L, Noujaim J (1998) Calcineurin and the biological effects of cyclosporine and tacrolimus. Transplant Proc 30:2167–2170
Hani J, Stumpf G, Domdey H (1995) PTF1 encodes an essential protein in *Saccharomyces cerevisiae*, which shows strong homology with a new putative family of PPIases. FEBS Lett 365:198–202
Hardwick JS, Kuruvilla FG, Tong JK, Shamji AF, Schreiber SL (1999) Rapamycin-modulated transcription defines the subset of nutrient-sensitive signaling pathways directly controlled by the Tor proteins. Proc Natl Acad Sci USA 96:14866–14870

Harrar Y, Bellini C, Faure JD (2001) FKBPs: at the crossroads of folding and transduction. Trends Plant Sci 6:426–431

Hassidim M, Schwarz R, Liemanhurwitz J, Marco E, Ronentarazi M, Kaplan A (1992) A cyanobacterial gene encoding peptidyl prolyl *cis/trans* isomerase. Plant Physiol 100:1982–1986

Heitman J, Movva NR, Hiestand PC, Hall MN (1991) FK 506-binding protein proline rotamase is a target for the immunosuppressive agent FK 506 in *Saccharomyces cerevisiae*. Proc Natl Acad Sci USA 88:1948–1952

Herrler M, Bang H, Marahiel MA (1994) Cloning and characterization of Ppib, a *Bacillus subtilis* gene which encodes a cyclosporin A-sensitive peptidyl prolyl *cis/trans* isomerase. Mol Microbiol 11:1073–1083

Hesterkamp T, Bukau B (1998) Role of the DnaK and HscA homologs of Hsp70 chaperones in protein folding in *E.coli*. EMBO J 17:4818–4828

Horne SM, Kottom TJ, Nolan LK, Young KD (1997) Decreased intracellular survival of an fkpA mutant of *Salmonella typhimurium* Copenhagen. Infect Immun 85:806–810

Horne SM, Young KD (1995) *Escherichia coli* and other species of the *Enterobacteriaceae* encode a protein similar to the family of Mip-like FK506-binding proteins. Arch Microbiol 163:357–365

Hsu T, McRackan D, Vincent TS, de Couet HG (2001) *Drosophila* Pin1 prolyl isomerase Dodo is a MAP kinase signal responder during oogenesis. Nat Cell Biol 3:538–543

Hsu VL, Handschumacher RE, Armitage IM (1990) Peptidyl prolyl *cis/trans* isomerase activity of cyclophilin studied by one-dimensional 1H nuclear magnetic resonance spectroscopy. J Am Chem Soc 112:6745–6747

Huai Q, Kim HY, Liu YD, Zhao YD, Mondragon A, Liu JO, Ke HM (2002) Crystal structure of calcineurin-cyclophilin-cyclosporin shows common but distinct recognition of immunophilin-drug complexes. Proc Natl Acad Sci USA 99:12037–12042

Huang HK, Forsburg SL, John UP, O'Connell MJ, Hunter T (2001) Isolation and characterization of the Pin1/Ess1p homolog in *Schizosaccharomyces pombe*. J Cell Sci 114:3779–3788

Hübner D, Drakenberg T, Forsen S, Fischer G (1991) Peptidyl prolyl *cis/trans* isomerase activity as studied by dynamic proton NMR spectroscopy. FEBS Lett 284:79–81

Huse M, Chen YG, Massague J, Kuriyan J (1999) Crystal structure of the cytoplasmic domain of the type I TGF-β receptor in complex with FKBP12. Cell 96:425–436

Janowski B, Wöllner S, Schutkowski M, Fischer G (1997) A protease free assay for peptidyl prolyl *cis/trans* isomerases using standard peptide substrates. Anal Biochem 252:299–307

Jiang HS, Xiong F, Kong SM, Ogawa T, Kobayashi M, Liu JO (1997) Distinct tissue and cellular distribution of two major isoforms of calcineurin. Mol Immunol 34:663–669

Jin JG, Melaragno MG, Liao DF, Yan C, Haendeler J, Suh YA, Lambeth JD, Berk BC (2000) Cyclophilin A is a secreted growth factor induced by oxidative stress. Circ Res 87:789–796

Jin L, Harrison SC, (2002) Crystal structure of human calcineurin complexed with cyclosporin A and human cyclophilin. Proc Natl Acad Sci USA 99:13522–13526

Joseph J D, Heitman J, Means AR (1999) Molecular cloning and characterization of Aspergillus nidulans cyclophilin B. Fungal Genet Biol 27:55–66

Kamphausen T, Fanghänel J, Neumann D, Schulz B, Rahfeld JU (2002) Characterization of *Arabidopsis thaliana At*FKBP42 that is membrane-bound and interacts with Hsp90. Plant J 32:263–276

Kandror O, Goldberg AL (1997) Trigger factor is induced upon cold shock and enhances viability of *Escherichia coli* at low temperatures. Proc Natl Acad Sci USA 94:4978–4981

Kern D, Drakenberg T, Wikstrom M, Forsen S, Bang H, Fischer G (1993) The *cis/trans* interconversion of the calcium regulating hormone calcitonin is catalyzed by cyclophilin. FEBS Lett 323:198–202

Kern D, Kern G, Scherer G, Fischer G, Drakenberg T (1995) Kinetic analysis of cyclophilin-catalyzed prolyl *cis/trans* isomerization by dynamic NMR spectroscopy. Biochemistry 34:13594–13602

Kern G, Kern D, Schmid FX, Fischer G (1994) Reassessment of the putative chaperone function of prolyl *cis/trans* isomerases. FEBS Lett 348:145–148

Kiefhaber T, Quaas R, Hahn U, Schmid FX (1990a) Folding of ribonuclease T1. 1. Existence of multiple unfolded states created by proline isomerization. Biochemistry 29:3053–3061

Kiefhaber T, Quaas R, Hahn U, Schmid FX (1990b) Folding of ribonuclease T1. 2. Kinetic models for the folding and unfolding reactions. Biochemistry 29:3061–3070

Kleerebezem M, Heutink M, Tommassen J (1995) Characterization of an *Escherichia coli* rotA mutant, affected in periplasmic peptidyl prolyl *cis/trans* isomerase. Mol Microbiol 18:313–320

Kofron JL, Kuzmic P, Kishore V, Colon Bonilla E, Rich DH (1991) Determination of kinetic constants for peptidyl prolyl *cis/trans* isomerases by an improved spectrophotometric assay. Biochemistry 30:6127–6134

Kok RG, Christoffels VM, Vosman B, Hellingwerf KJ (1994) A gene of *Acinetobacter calcoaceticus* BD413 encodes a periplasmic peptidyl prolyl *cis/trans* isomerase of the cyclophilin sub-class that is not essential for growth. Biochim Biophys Acta 1219:601–606
Kontinen VP, Sarvas M (1993) The PrsA lipoprotein is essential for protein secretion in *Bacillus subtilis* and sets a limit for high-level secretion. Mol Microbiol 8:727–737
Kramer ML, Fischer G (1997) FKBP-like catalysis of peptidyl prolyl bond isomerization by micelles and membranes. Biopolymers 42:49–60
Küllertz G, Lüthe S, Fischer G (1998) Semiautomated microtiter plate assay for monitoring peptidyl prolyl *cis/trans* isomerase activity in normal and pathological human sera. Clin Chem 44:502–508
Kung L, Halloran PF (2002) Immunophilins may limit calcineurin inhibition by cyclosporine and tacrolimus at high drug concentrations. Transplantation 70:327–335
Kurek I, Aviezer K, Erel N, Herman E, Breiman A (1999) The wheat peptidyl prolyl *cis/trans* isomerase FKBP77 is heat induced and developmentally regulated. Plant Physiol 119:693–703
Kurek I, Pirkl F, Fischer E, Buchner J, Breiman A (2002) Wheat FKBP73 functions in vitro as a molecular chaperone independently of its peptidyl prolyl *cis/trans* isomerase activity. Planta 215:119–126
Lang K, Schmid FX, Fischer G (1987) Catalysis of protein folding by prolyl isomerase. Nature 329:268–70
Langer T, Lu C, Echols H, Flanagan J, Hayer MK, Hartl FU (1992) Successive action of DnaK, DnaJ and GroEL along the pathway of chaperone-mediated protein folding. Nature 356:683–689
Lee JP, Palfrey HC, Bindokas VP, Ghadge GD, Ma L, Miller RJ, Roos RP (1999) The role of immunophilins in mutant superoxide dismutase-1-linked familial amyotrophic lateral sclerosis. Proc Natl Acad Sci USA 96:3251–3256
Leiva MC, Lyttle CR (1992) Leukocyte chemotactic activity of FKBP and inhibition by FK506. Biochem Biophys Res Commun 186:1178–1183
Leverson JD, Ness SA (1998) Point mutations in v-myb disrupt cyclophilin-catalyzed negative regulatory mechanism. Mol Cell 1:203–211
Li HB, Oberhauser AF, Redick SD, Carrion-Vazquez M, Erickson HP, Fernandez JM (2001a) Multiple conformations of PEVK proteins detected by single-molecule techniques. Proc Natl Acad Sci USA 98:10682–10686
Li TK, Baksh S, Cristillo AD, Bierer BE (2002) Calcium- and FK506-independent interaction between the immunophilin FKBP51 and calcineurin. J Cell Biochem 84:460–471
Li ZY, Liu CP, Zhu LQ, Jing GZ, Zhou JM (2001b) The chaperone activity of trigger factor is distinct from its isomerase activity during coexpression with adenylate kinase in *Escherichia coli*. FEBS Lett 506:108–112
Lin DT, Lechleiter JD (2002) Mitochondrial targeted cyclophilin D protects cells from cell death by peptidyl prolyl isomerization. J Biol Chem 277:31134–31141
Lin LN, Brandts JF (1979) Evidence suggesting that some proteolytic enzymes may cleave only the trans form of the peptide bond. Biochemistry 18:43–47
Lin LN, Brandts JF (1985) Isomer-specific proteolysis of model substrates: influence that the location of the proline residue exerts on *cis/trans* specificity. Biochemistry 24:6533–6538
Lin LN, Hasumi H, Brandts JF (1988) Catalysis of proline isomerization during protein-folding reactions. Biochim Biophys Acta 956:256–266
Liou YC, Ryo A, Huang HK, Lu PJ, Bronson R, Fujimori F, Uchida T, Hunter T, Lu KP (2002) Loss of Pin1 function in the mouse causes phenotypes resembling cyclin D1-null phenotypes. Proc Natl Acad Sci USA 99:1335–1340
Liu J, Farmer JD, Jr. Lane WS, Friedman J, Weissman I, Schreiber SL (1991) Calcineurin is a common target of cyclophilin-cyclosporin A and FKBP-FK506 complexes. Cell 66:807–815
London RE, Davis DG, Vavrek RJ, Stewart JM, Handschumacher RE (1990) Bradykinin and its Gly6 analog are substrates of cyclophilin: a fluorine-19 magnetization transfer study. Biochemistry 29:10298–10302
Lopezilasaca M, Schiene C, Küllertz G, Tradler T, Fischer G, Wetzker R (1998) Effects of FK506 binding protein 12 and FK506 on autophosphorylation of epidermal growth factor receptor. J Biol Chem 273:9430–9434
Lu KP, Liou YC, Zhou XZ (2002) Pinning down proline-directed phosphorylation signaling. Trends Cell Biol 12:164–172
Lu PJ, Zhou XZ, Shen MH, Lu KP (1999) Function of WW domains as phosphoserine- or phosphothreonine-binding modules. Science 283:1325–1328
Luban J (1996) Absconding with the chaperone—essential cyclophilin gag interaction in HIV-1 virions. Cell 87:1157–1159

Lyon WR, Gibson CM, Caparon MG (1998) A role for trigger factor and rgg-like regulator in the transcription, secretion and processing of the cysteine proteinase of *Streptococcus pyogenes*. EMBO J 17:6263–6275

MacArthur MW, Thornton JM (1991) Influence of proline residues on protein conformation. J Mol Biol 218:397–412

Maier R, Scholz C, Schmid FX (2001) Dynamic association of trigger factor with protein substrates. J Mol Biol 314:1181–1190

Maleszka R, Hanes SD, Hackett RL, deCouet HG, Miklos GLG (1996) The *Drosophila melanogaster* dodo (dod) gene, conserved in humans, is functionally interchangeable with the ESS1 cell division gene of *Saccharomyces*. Proc Natl Acad Sci USA 93:447–451

Mallis RJ, Brazin KN, Fulton DB, Andreotti AH (2002) Structural characterization of a proline-driven conformational switch within the Itk SH2 domain. Nat Struct Biol 9:900–905

Mamane Y, Sharma S, Petropoulos L, Lin RT, Hiscott J (2000) Posttranslational regulation of IRF-4 activity by the immunophilin FKBP52. Immunity 12:129–140

Marivet J, Frendo P, Burkard G (1992) Effects of abiotic stresses on cyclophilin gene expression in maize and bean and sequence analysis of bean cyclophilin cDNA. Plant Sci 84:171–178

Mayr LM, Odefey C, Schutkowski M, Schmid FX (1996) Kinetic analysis of the unfolding and refolding of ribonuclease T1 by stopped-flow double-mixing technique. Biochemistry 35:5550–5561

Mayr LM, Schmid FX (1993) Kinetic models for unfolding and refolding of ribonuclease T1 with substitution of *cis*-proline 39 by alanine. J Mol Biol 231:913–926

Mayr LM, Willbold D, Rosch P, Schmid FX (1994) Generation of a nonprolyl *cis* peptide bond in ribonuclease T1. J Mol Biol 240:288–293

McCarthy AA, Haebel PW, Torronen A, Rybin V, Baker EN, Metcalf P (2000) Crystal structure of the protein disulfide bond isomerase, DsbC, from *Escherichia coli*. Nat Struct Biol 7:196–199

McCornack MA, Kakalis LT, Caserta C, Handschumacher RE, Armitage IM (1997) HIV protease substrate conformation—modulation by cyclophilin A. FEBS Lett 414:84–88

Meng X, Lu XJ, Morris CA, Keating MT (1998) Novel human gene FKBP6 is deleted in Williams-syndrome. Genomics 52:130–137

Merker MP, Dawson CA (1995) Cyclophilin-facilitated bradykinin inactivation in the perfused rat lung. Biochem Pharmacol 50:2085–2091

Merker MP, Dawson CA, Bongard RD, Roerig DL, Haworth ST, Linehan JH (1993) Angiotensin-converting enzyme preferentially hydrolyzes *trans* isomer of proline-containing substrate. J Appl Physiol 75:1519–1524

Meyer S, Jabs A, Schutkowski M, Fischer G (1994) Separation of *cis/trans* isomers of a prolyl peptide bond by capillary zone electrophoresis. Electrophoresis 15:1151–1157

Mihatsch MJ, Kyo M, Morozumi K, Yamaguchi Y, Nickeleit V, Ryffel B (1998) The side effects of cyclosporin A and tacrolimus. Clin Nephrol 49:356–363

Montague JW, Gaido ML, Frye C, Cidlowski JA (1994) A calcium-dependent nuclease from apoptotic rat thymocytes is homologous with cyclophilin-recombinant cyclophilins A, B, and C have nuclease activity. J Biol Chem 269:18877–18880

Moritz R, Reinstadler D, Fabian H, Naumann D (2002) Time-resolved FTIR difference spectroscopy as tool for investigating refolding reactions of ribonuclease T1 synchronized with *trans* to *cis* prolyl isomerization. Biopolymers 67:145–155

Mücke M, Schmid FX (1992) Enzymatic catalysis of prolyl isomerization in an unfolding protein. Biochemistry 31:7848–7854

Navia MA (1996) Protein-drug complexes important for immunoregulation and organ transplantation. Curr Opin Struct Biol 6:838–847

Odefey C, Mayr LM, Schmid FX (1995) Nonprolyl *cis/trans* peptide bond isomerization as a rate-determining step in protein unfolding and refolding. J Mol Biol 245:69–78

Ohta T, Ishimura K, Takitanis S (1993) Selective online enrichment and separation of peptides having aromatic amino acids at their C-termini by column switching using an anhydrochymotrypsin-immobilized precolumn. J Chromatogr 637:35–41

Olson ST, Bock PE, Kvassman J, Shore JD, Lawrence DA, Ginsburg D, Bjork I (1995) Role of the catalytic serine in the interactions of serine proteinases with protein inhibitors of the serpin family. Contribution of a covalent interaction to the binding energy of serpin-proteinase complexes. J Biol Chem 270:30007–30017

Ou WB, Luo W, Park YD, Zhou HM (2000) Chaperone-like activity of peptidyl prolyl *cis/trans* isomerase during creatine kinase refolding. Protein Sci 10:2346–2353

Paek KH, Walker GC (1987) *Escherichia coli* DnaK null mutants are inviable at high temperature. J Bacteriol 169:283–290

Pappenberger G, Aygun H, Engels JW, Reimer U, Fischer G, Kiefhaber T (2001) Nonprolyl *cis* peptide bonds in unfolded proteins cause complex folding kinetics. Nat Struct Biol 8:452–458

Patzelt H, Rüdiger S, Brehmer D, Kramer G, Vorderwulbecke S, Schaffitzel E, Waitz A, Hesterkamp T, Dong L, Schneider-Mergener J, Bukau B (2001) Binding specificity of *Escherichia coli* trigger factor. Proc Natl Acad Sci USA 98:14244–14249

Pereira PJB, Vega MC, Gonzalez-Rey E, Fernandez-Carazo R, Macedo-Ribeiro S, Gomis-Ruth FX, Gonzalez A, Coll M (2002) *Trypanosoma cruzi* macrophage infectivity potentiator has a rotamase core and a highly exposed α-helix. EMBO Rep 3:88–94

Perez-Perez JM, Ponce MR, Micol JL (2001) Mutations in the ULTRACURVATA2 gene of *Arabidopsis thaliana*, which encodes a FKBP-like protein, cause dwarfism, leaf epinasty and helical rotation of several organs. Int J Dev Biol 45(S1): 49–50

Pirkl F, Fischer E, Modrow S, Buchner J (2001) Localization of the chaperone domain of FKBP52. J Biol Chem 276:37034–37041

Pissavin C, Hugouvieuxcottepattat N (1997) Characterization of a periplasmic peptidyl prolyl *cis/trans* isomerase in *Erwinia chrysanthemi*. FEMS Microbiol Lett 157:59–65

Pushkarsky T, Zybarth G, Dubrovsky L, Yurchenko V Tang H, Guo HM, Toole B, Sherry B, Bukrinsky M (2001) CD147 facilitates HIV-1 infection by interacting with virus-associated cyclophilin A. Proc Natl Acad Sci USA 98:6360–6365

Ramm K, Plückthun A (2000) The periplasmic *Escherichia coli* peptidyl prolyl *cis/trans* isomerase FkpA— II Isomerase-independent chaperone activity in vitro. J Mol Biol 310:485–498

Ramm K, Plückthun, A (2001) High enzymatic activity and chaperone function are mechanistically related features of the dimeric *E. coli* peptidyl prolyl isomerase FkpA. J Biol Chem 275:17106–17113

Ranganathan R, Lu KP, Hunter T, Noel JP (1997) Structural and functional analysis of the mitotic rotamase Pin1 suggests substrate recognition is phosphorylation. Cell 89:875–886

Reimer U, Fischer G (2002) Local structural changes caused by peptidyl prolyl *cis/trans* isomerization in the native state of proteins. Biophys Chem 96:203–212

Reimer U, Scherer G, Drewello M, Kruber S, Schutkowski M, Fischer G (1998) Side-chain effects on peptidyl prolyl *cis/trans* isomerization. J Mol Biol 279:449–460

Reyes DY, Yoshikawa H (2002) DnaK chaperone machine and trigger factor are only partially required for normal growth of *Bacillus subtilis*. Biosci Biotechnol Biochem 66:1583–1586

Riboldi-Tunnicliffe A, König B, Jessen S, Weiss MS, Rahfeld J, Hacker J, Fischer G, Hilgenfeld R (2001) Crystal structure of Mip, a prolylisomerase from *Legionella pneumophila*. Nat Struct Biol 8:779–783

Rippmann JF, Hobbie S, Daiber C, Guilliard B, Bauer M, Birk J, Nar H, Garin-Chesa P, Rettig WJ, Schnapp A (2000) Phosphorylation-dependent proline isomerization catalyzed by Pin1 is essential for tumor cell survival and entry into mitosis. Cell Growth Differ 11:409–416

Rizzitello AE, Harper JR, Silhavy TJ (2001) Genetic evidence for parallel pathways of chaperone activity in the periplasm of *Escherichia coli*. J Bacteriol 183:6794–6800

Robson T, Joiner MC, Wilson GD, McCullough W, Price ME, Logan I, Jones H, McKeown SR, Hirst DG (1999) A novel human stress response-related gene with a potential role in induced radioresistance. Radiat Res 152:451–461

Roof WD, Horne SM, Young KD, Young R (1994) SlyD, a host gene required for phi X174 lysis, is related to the FK506-binding protein family of peptidyl prolyl *cis/trans* isomerases. J Biol Chem 269:2902–2910

Rycyzyn MA, Clevenger CV (2002) The intranuclear prolactin/cyclophilin B complex as a transcriptional inducer. Proc Natl Acad Sci USA 99:6790–6795

Rycyzyn MA, Reilly SC, O'Malley K, Clevenger CV (2000) Role of cyclophilin B in prolactin signal transduction and nuclear retrotranslocation. Mol Endocrinol 14:1175–1186

Sabers CJ, Martin MM, Brunn GJ, Williams JM, Dumont FJ, Wiederrecht G, Abraham RT (1995) Isolation of a protein target of the FKBP12-rapamycin complex in mammalian cells. J Biol Chem 270:815–822

Saphire ACS, Bobardt MD, Gallay PA (2002) Transcomplementation rescue of cyclophilin A-deficient viruses reveals that the requirement for cyclophilin A in human immunodeficiency virus type 1. J Virol 76:2255–2262

Schaffitzel E, Rüdiger S, Bukau B, Deuerling E (2001) Functional dissection of trigger factor and DnaK: Interactions with nascent polypeptides and thermally denatured proteins. Biol Chem 382:1235–1243

Scherer G, Kramer ML, Schutkowski M, Reimer U, Fischer G (1998) Barriers to rotation of secondary amide peptide bonds. J Amer Chem Soc 120:5568–5574

Schiene C, Reimer U, Schutkowski M, Fischer G (1998) Mapping the stereospecificity of peptidyl prolyl *cis/trans* isomerases. FEBS Lett 432:202–206

Schiene-Fischer C, Fischer G (2001) Direct measurement indicates a slow *cis/trans* isomerization at the secondary amide peptide bond of glycylglycine. J Amer Chem Soc 123:6227–6231

Schiene-Fischer C, Habazettl J, Schmid FX, Fischer G (2002a) The hsp70 chaperone DnaK is a secondary amide peptide bond *cis/trans* isomerase. Nat Struct Biol 9:419–424

Schiene-Fischer C, Habazettl J, Tradler T, Fischer G (2002b) Evaluation of similarities in the *cis/trans* isomerase function of trigger factor and DnaK. Biol Chem 383:1865–1875

Schiene-Fischer C, Yu C (2001) Receptor accessory folding helper enzymes: the functional role of peptidyl prolyl *cis/trans* isomerases. FEBS Lett 495:1-6

Schmid FX (1986) Fast-folding and slow-folding forms of unfolded proteins. Methods Enzymol 131:70–82

Schmidt B, Tradler T, Rahfeld JU, Ludwig B, Jain B, Mann K, Rücknagel KP, Janowski B, Schierhorn A, Küllertz G, Hacker J, Fischer G (1996) A cyclophilin-like peptidyl prolyl *cis/trans* isomerases from *Legionella pneumophila*-characterization, molecular cloning, and overexpression. Mol Microbiol 21:1147–1160

Scholz C, Mücke M, Rape M, Pecht A, Pahl A, Bang H, Schmid FX (1998a) Recognition of protein substrates by the prolyl isomerase trigger factor is independent of proline residues. J Mol Biol 277:723–732

Scholz C, Scherer G, Mayr LM, Schindler T, Fischer G, Schmid FX (1998b) Prolyl isomerases do not catalyze isomerization of nonprolyl peptide bonds. Biol Chem 379:361–365

Scholz C, Stoller G, Zarnt T, Fischer G, Schmid FX (1997) Cooperation of enzymatic and chaperone functions of trigger factor in the catalysis of protein folding. EMBO J 16:54–58

Schreiber SL, Crabtree GR (1992) The mechanism of action of cyclosporin-A and FK506. Immunol Today 13:136–142

Schreier MH, Baumann G, Zenke G (1993) Inhibition of T-Cell signaling pathways by immunophilin drug complexes are side effects inherent to immunosuppressive properties. Transplant Proc 25:502–507

Schutkowski M, Drewello M, Wöllner S, Jakob M, Reimer U, Scherer G, Schierhorn A, Fischer G (1996b) Extended binding sites of cyclophilin as revealed by HIV-1 gag polyprotein-derived oligopeptides. FEBS Lett 394:289–294

Schutkowski M, Wöllner S, Fischer G (1996a) Inhibition of peptidyl prolyl *cis/trans* isomerase activity by substrate analog structures: Thioxo tetrapeptide-4-nitroanilides. Biochemistry 34:13016–13026

Sekerina E, Rahfeld JU, Müller J, Fanghänel J, Rascher C, Fischer G, Bayer P (2000) NMR solution structure of hPar14 reveals similarity to the peptidyl prolyl *cis/trans* isomerase domain of the mitotic regulator hPin1 but indicates a different functionality of the protein. J Mol Biol 301:1003–1017

Sharma S, Mamane Y, Grandvaux N, Bartlett J, Petropoulos L, Lin RT, Hiscott J (2000) Activation and regulation of interferon regulatory factor 4 in HTLV type 1-infected T lymphocytes. AIDS Res Hum Retroviruses 16:1613–1622

Shaw PE (2002) Peptidyl prolyl isomerases: a new twist to transcription EMBO Rep 3:521–526

Sherman MY, Goldberg AL (1992) Involvement of the chaperonin DnaK in the rapid degradation of a mutant protein in *Escherichia coli.* EMBO J 11:71–77

Shou WN, Aghdasi B, Armstrong DL, Guo QX, Bao SD, Charng MJ, Mathews LM, Schneider MD, Hamilton SL, Matzuk MM (1998) Cardiac defects and altered ryanodine receptor function in mice lacking FKBP12. Nature 391:489–492

Sigal NH, Dumont F, Durette P, Siekierka JJ, Peterson L, Rich DH, Dunlap BE, Staruch MJ, MelinoMR, Koprak SL, Willimas D, Witzel B, Pisano JM (1991) Is cyclophilin involved in the immunosuppressive and nephrotoxic mechanism of action of cyclosporin A? J Exp Med 173:619–628

Somers PK, Wandless TJ, Schreiber SL (1991) Synthesis and analysis of 506BD, a high affinity ligand for the immunophilin FKBP. J Am Chem Soc 113:8045–8056

Spiess C, Beil A, Ehrmann M (1999) A temperature-dependent switch from chaperone to protease in a widely conserved heat shock protein. Cell 97:339–347

Steiner JP, Connolly MA, Valentine HI, Hamilton GS, Dawson TM, Hester L, Snyder SH (1997) Neurotrophic actions of nonimmunosuppressive analogs of immunosuppressive drugs FK506, rapamycin and cyclosporine. Nat Med 3:421–428

Stukenberg PT, Kirschner MW (2001) Pin1 acts catalytically to promote a conformational change in Cdc25. Mol Cell 7:1071–1083

Sydenham M, Douce G, Bowe F, Ahmed S, Chatfield S, Dougan G (2000) *Salmonella enterica serovar Typhimurium* surA mutants are attenuated and effective live oral vaccines. Infect Immun 68:1109–1115

Sykes K, Gething MJ, Sambrook J (1993) Proline isomerases function during heat shock. Proc Natl Acad Sci USA 90:5853–5857

Tang WX, Chen YF, Zou AP, Campbell WB, Li PL (2002) Role of FKBP12.6 in cADPR-induced activation of reconstituted ryanodine receptors from arterial smooth muscle. Am J Physiol 282: H1304-H1310

Teter SA, Houry WA, Ang D, Tradler T, Rockabrand D, Fischer G, Blum P, Georgopoulos C, Hartl FU (1999) Polypeptide flux through bacterial Hsp70: DnaK cooperates with trigger factor in chaperoning nascent chains. Cell 97:755–765

Thomson JA, Shirley BA, Grimsley GR, Pace CN (1989) Conformational stability and mechanism of folding of ribonuclease T1. J Biol Chem 264:11614–11620

Thunecke F, Fischer G (1998) Separation of *cis/trans* conformers of human and salmon calcitonin by low temperature capillary electrophoresis. Electrophoresis 19:288–294

Tormo A, Almiron M, Kolter R (1990) SurA, an *Escherichia coli* gene essential for survival in stationary phase. J Bacteriol 172:4339–4347

Truffa-Bachi P, Lefkovits I, Frey JR (2000) Proteomic analysis of T cell activation in the presence of cyclosporin A: immunosuppressor and activator removal induces de novo protein synthesis. Mol Immunol 37:21–28

Tsai CJ, Ma B, Nussinov R (1999) Folding and binding cascades: shifts in energy landscapes. Proc Natl Acad Sci USA 96:9970–9972

Van Melderen L, Gottesman S (1999) Substrate sequestration by a proteolytically inactive Lon mutant. Proc Natl Acad Sci USA 96:6064–6071

Vance JE, Leblanc DA, London RE (1997b) Cleavage of the X-Pro peptide bond by pepsin is specific for the *trans* isomer. Biochemistry 36:13232–13240

Vance JE, Leblanc DA, Wingfield P, London RE (1997a) Conformational selectivity of HIV-1 protease cleavage of Xaa-Pro bonds and its implications. J Biol Chem 272:15603–15606

Videen JS, Stamnes MA, Hsu VL, Goodman M (1994) Thermodynamics of cyclophilin catalyzed peptidyl prolyl isomerization by NMR spectroscopy. Biopolymers 34:171–175

Vittorioso P, Cowling R, Faure JD, Caboche M, Bellini C (1998) Mutations in the *Arabidopsis paticcino*1 gene, which encodes a new FK506-binding protein-like protein, has a dramatic effect on plant development. Mol Cell Biol 18:3034–3043

Vogel KW, Briesewitz R, Wandless TJ, Crabtree GR (2001) Calcineurin inhibitors and the generalization of the presenting protein strategy. Adv Protein Chem 56:253–291

von Ahsen O, Lim JH, Caspers P, Martin F, Schönfeld HJ, Rassow J, Pfanner N (2000) Cyclophilin-promoted folding of mouse dihydrofolate reductase does not include the slow conversion of the late-folding intermediate to the active enzyme. J Mol Biol 294:809–818

Waldmeier PC, Feldtrauer JJ, Qian T, Lemasters JJ (2002) Inhibition of the mitochondrial permeability transition by the nonimmunosuppressive cyclosporin derivative NIM811. Mol Pharmacol 62:22–29

Wang P, Cardenas ME, Cox CM, Perfect JR, Heitman J (2001) Two cyclophilin A homologs with shared and distinct functions important for growth and virulence of *Cryptococcus neoformans*. EMBO Rep 2:511–518

Webb HM, Ruddock LW, Marchant RJ, Jonas K, Klappa P (2001) Interaction of the periplasmic peptidyl prolyl *cis/trans* isomerase SurA with model peptides—The N-terminal region of SurA is essential and sufficient for peptide binding. J Biol Chem 276:45622–45627

Webel R, Menon I, O'Tousa JE, Colley NJ (2000) Role of asparagine-linked oligosaccharides in rhodopsin maturation and association with its molecular chaperone, NinaA. J Biol Chem 275:24752–24759

Weisman R, Finkelstein S, Choder M (2001) Rapamycin blocks sexual development in fission yeast through inhibition of the cellular function of an FKBP12 homolog. J Biol Chem 276:24736–24742

Weisshoff H, Frost K, Brandt W, Henklein P, Mügge C, Frommel C (1995) Novel disulfide-constrained pentapeptides as models for β-VIa turns in proteins. FEBS Lett 372:203–209

Weiwad M, Küllertz G, Schutkowski M, Fischer G (2000) Evidence that the substrate backbone conformation is critical to phosphorylation by p42 MAP kinase. FEBS Lett 478:39–42

Wiederrecht G, Brizuela L, Elliston K, Sigal NH, Siekierka JJ (1991) FKB1 encodes a nonessential FK 506-binding protein in *Saccharomyces cerevisiae* and contains regions suggesting homology to the cyclophilins. Proc Natl Acad Sci USA 88:1029–1033

Winkler KE, Swenson KI, Kornbluth S, Means AR (2000) Requirement of the prolyl isomerase Pin1 for the replication checkpoint. Science 287:1644–1647

Wintermeyer E, Ludwig B, Steinert M, Schmidt B, Fischer G, Hacker J (1995) Influence of site specifically altered Mip proteins on intracellular survival of *Legionella pneumophila* in eukaryotic cells. Infect Immun 63:4576–4583

Witte A, Schrot G, Schon P, Lubitz W (1997) Proline 21, a residue within the α-helical domain of phiX174 lysis protein E, is required for its function in *Escherichia coli*. Mol Microbiol 26:337–346

Wong CYF, Heuzenroeder MW, Quinn DM, Flower RLP (1997) Cloning and characterization of two immunophilin-like genes IlpA and FkbpA on a single 3.9-kilobase fragment of *Aeromonas hydrophila* genomic DNA. J Bacteriol 179: 3397–3403

Wu XY, Wilcox CB, Devasahayam G, Hackett RL, Arevalo-Rodriguez M, Cardenas ME, Heitman J, Hanes SD (2000) The Ess1 prolyl isomerase is linked to chromatin remodeling complexes and the general transcription machinery. EMBO J 19:3727–3738

Wüthrich K, Grathwohl C (1974) A novel approach for studies of the molecular conformations in flexible polypeptides. FEBS Lett 43:337–340

Xin HB, Senbonmatsu T, Cheng DS, Wang YX, Copello JA, Ji GJ, Collier ML, Deng KY, Jeyakumar LH, Magnuson MA, Inagami T, Kotlikoff MI, Fleischer S (2002) Estrogen protects FKBP12.6 null mice from cardiac hypertrophy. Nature 416:334–337

Xu Q, Leiva MC, Fischkoff SA, Handschumacher RE, Lyttle CR (1992) Leukocyte chemotactic activity of cyclophilin. J Biol Chem 267:11968–11971

Yaffe MB, Schutkowski M, Shen MH, Zhou XZ, Stukenberg PT, Rahfeld JU, Xu J, Kuang J, Kirschner MW, Fischer G, Cantley LC, Lu KP (1997) Sequence-specific and phosphorylation-dependent proline isomerization: A potential mitotic regulatory mechanism. Science 278:1957–1960

Yang WM, Yao YL, Seto E (2001) The FK506-binding protein 25 functionally associates with histone deacetylases and with transcription factor YY1. EMBO J 20:4814–4825

Yao J, Feher VA, Espejo BF, Reymond MT, Wright PE, Dyson HJ (1994) Stabilization of a type VI turn in a family of linear peptides in water solution. J Mol Biol 243:736–753

Yoo SH, Myszka DG, Yeh CY, McMurray M, Hill CP, Sundquist WI (1997) Molecular recognition in the HIV-1 capsid/cyclophilin A complex. J Mol Biol 269: 780–795

Yuan XM, Werner JM, Knott V, Handford PA, Campbell ID, Downing AK (1998) Effects of proline *cis/trans* isomerization on TB domain secondary structure. Protein Sci 7:2127–2135

Yurchenko V, O'Connor M, Dai WW, Guo HM, Toole B, Sherry B, Bukrinsky M (2001) CD147 is a signaling receptor for cyclophilin B. Biochem Biophys Res Commun 288:786–788

Yurchenko V, Zybarth G, O'Connor M, Dai WW, Franchin G, Hao T, Guo HM, Hung HC, Toole B, Gallay P, Sherry B, Bukrinsky M (2002) Active site residues of cyclophilin A are crucial for its signaling activity via CD147. J Biol Chem 277:22959–22965

Zenke G, Strittmatter U, Fuchs S, Quesniaux VFJ, Brinkmann V, Schuler W, Zurini M, Enz A, Billich A, Sanglier JJ, Fehr T (2001) Sanglifehrin A, a novel cyclophilin-binding compound showing immunosuppressive activity with a new mechanism of action. J Immunol 166:7165–7171

Zhang W, Zimmer G, Chen JZ, Ladd D, Li E, Alt FW, Wiederrecht G, Cryan J, Oneill EA, Seidman CE. Abbas AK, Seidman JG (1996) T cell response in calcineurin A α-deficient mice. J Exp Med 183:413–420

Zhang YX, Fussel S, Reimer U, Schutkowski M, Fischer G (2002) Substrate-based design of reversible Pin1 inhibitors. Biochemistry 41:11868–11877

Zhao YD, Chen YQ, Schutkowski M, Fischer G, Ke HM (1997) Cyclophilin A complexed with a fragment of HIV-1 gag protein—insights into HIV-1 infectious activity. Structure 5:139–146

Zhou XZ, Kops O, Werner A, Lu PJ, Shen MH, Stoller G, Küllertz G, Stark M, Fischer G, Lu KP (2000) Pin1-dependent prolyl isomerization regulates dephosphorylation of Cdc25C and τproteins. Mol Cell 6:873–883

Instructions for authors

1 Legal requirements

The author(s) guarantee(s) that the manuscript will not be published elsewhere in any language without the consent of the copyright holders, that the rights of third parties will not be violated, and that the publisher will not be held legally responsible should there be any claims for compensation.

Authors wishing to include figures or text passages that have already been published elsewhere are required to obtain permission from the copyright holder(s) and to include evidence that such permission has been granted when submitting their papers. Any material received without such evidence will be assumed to originate from the authors.

Manuscripts must be accompanied by the "Copyright Transfer Statement".

Please include at the end of the acknowledgements a declaration that the experiments comply with the current laws of the country in which they were performed.

2 Editorial procedure

Manuscripts should be submitted in English, together with one set of illustrations and a complete pdf file, to the editor in charge.

The author is responsible for the accuracy of the references.

3 Manuscript preparation

To help you prepare your manuscript, Springer offers a template that can be used with Winword 7 (Windows 95), Winword 6 and Word for Macintosh.

For details see point 4.

All manuscripts are subject to copy editing.

■ **Title page**

- The name(s) of the author(s)
- A concise and informative title
- The affiliation(s) and address(es) of the author(s)
- The e-mail address, telephone and fax numbers of the communicating author

■ **Abstract.** Each paper must be preceded by an abstract presenting the most important results and conclusions.

■ **Abbreviations** should be defined at first mention in the abstract and again in the main body of the text and used consistently thereafter

A list of **symbols** should follow the abstract if such a list is needed. Symbols must be written clearly. The international system of units (SI units) should be used. The numbering of chapters should be in decimal form.

Footnotes on the title page are not given reference symbols. Footnotes to the text are numbered consecutively; those to tables should be indicated by superscript lower-case letters (or asterisks for significance values and other statistical data).

Acknowledgements. These should be as brief as possible. Any grant that requires acknowledgement should be mentioned The names of funding organizations should be written in full.

Funding. Authors are expected to disclose any commercial or other associations that might pose a conflict of interest in connection with submitted material. All funding sources supporting the work and institutional or corporate affiliations of the authors should be acknowledged.

■ **References**

The list of References should only include works that are cited in the text and that have been published or accepted for publication. Personal communications should only be mentioned in the text.

In the text, references should be cited by author and year (e.g. Hammer 1994; Hammer and Sjöqvist 1995; Hammer et al. 1993) and listed in alphabetical order in the reference list.

Examples:

Monographs:
Snider T, Grand L (1982) Air pollution by nitrogen oxides. Elsevier, Amsterdam

Anthologies and proceedings:
Noller C, Smith VR (1997) Ultraviolet selection pressure on earliest organisms. In: Kingston H, Fulling CP (eds) Natual environment background analysis. Oxford University Press, Oxford, pp 211–219

Journals:
Meltzoff AN, Moore MK (1977) Imitation of facial and manual gestures by human neonates. Science 198:75–78

If available the Digital Object Identifier (DOI) of the cited literature should be added at the end of the reference in question.

■ **Illustrations and Tables**

All figures (photographs, graphs or diagrams) and tables should be cited in the text, and each numbered consecutively throughout. Figure parts should be identified by lower-case roman letters. The placement of figures and tables should be indicated in the left margin. For submission of figures in electronic form see below

Line drawings. Please submit good-quality prints. The inscriptions should be clearly legible.

Half-tone illustrations (black and white and color). Please submit well-contrasted photographic prints with the top indicated on the back. Magnification should be indicated by scale bars.

Figure legends must be brief, self-sufficient explanations of the illustrations. The legends should be placed at the end of the text.

Tables should have a title and a legend explaining any abbreviation used in that table. Footnotes to tables should be indicated by superscript lower-case letters (or asterisks) for significance values and other statistical data.

4 Electronic submission of final version

Please send only the final version of the article, as accepted by the editors.

Preparing your manuscript

The template is available:

→ via ftp:
Address: ftp.springer.de/
User ID: ftp
Password: your own e-mail address
- Directory: /pub/Word/journals
- File names: either sv-journ.zip or sv-journ.doc and sv-journ.dot

→ via browser
- http://www.springer.de/author/index.html

The zip file should be sent uuencoded.

Layout guidelines

1. Use a normal, plain font (e.g., Times Roman) for text.
   Other style options:
   - for textual emphasis use italic types.
   - for special purposes, such as for mathematical vectors, use boldface type.
2. Use the automatic page numbering function to number the pages.
3. Do not use field functions.
4. For indents use tab stops or other commands, not the space bar.
5. Use the table functions of your word processing program, not spreadsheets, to make tables.
6. Use the equation editor of your word processing program or MathType for equations.
7. Place any figure legends or tables at the end of the manuscript.
8. Submit all figures as separate files and do not integrate them within the text.

Data formats

Save your file in two different formats:

1. RTF (Rich Text Format) or Word compatible Word 95/97
2. pdf (a single pdf file including text, tables and figures)

Illustrations

The preferred figure formats are EPS for vector graphics exported from a drawing program and TIFF for halftone illustrations. EPS files must always contain a preview in TIFF of the figure. The file name (one file for each figure) should include the figure number. Figure legends should be included in the text and not in the figure file.

Scan resolution: Scanned line drawings should be digitized with a minimum resolution of 800 dpi relative to the final figure size. For digital halftones, 300 dpi is usually sufficient.

Color illustrations: Store color illustrations as RGB (8 bits per channel) in TIFF format.

General information on data delivery

Please send us a zip file (text and illustrations in separate files) either:

→ Via ftp.springer.de
(to our ftp.server; log-in "anonymous"; password: your e-mail address; further information in the readme file on the server)

→ By e-mail
(only suitable for small volumes of data)

→ or on any of the following media:
- On a diskette [you may use .tar, .zip, .gzip (.gz), .sit, and compress (.Z)]
- On a ZIP cartridge
- On a CD-ROM

Please always supply the following information with your data: journal title, operating system, word processing program, drawing program, image processing program, compression program.

The file name should be memorable (e.g., author name), have no more than 8 characters, and include no accents or special symbols. Use only the extensions that the program assigns automatically.

5 Proofreading

Authors should make their proof corrections on a printout of the pdf file supplied, checking that the text is complete and that all figures and tables are included. After online publication, further changes can only be made in the form of an Erratum, which will be hyperlinked to the article. The author is entitled to formal corrections only. Substantial changes in content, e.g. new results, corrected values, title and authorship are not allowed without the approval of the editor in charge. In such a case please contact the Editor in charge before returning the proofs to the publisher.

6 Offprints, Free copy

You are entitled to receive a pdf file of your article for your own personal use. Orders for offprints can be placed by returning the order form with the corrected proofs. One complimentary copy of the issue in which your article appears is supplied.

Printing: Saladruck Berlin
Binding: Stürtz AG, Würzburg

www.ingramcontent.com/pod-product-compliance
Lightning Source LLC
Chambersburg PA
CBHW071234230926
51849CB00001B/95
* 9 7 8 3 5 4 0 4 0 1 3 6 0 *